APPLIED SOCIOLOGY
TERMS, TOPICS, TOOLS, AND TASKS

Stephen F. Steele
Anne Arundel Community College

Jammie Price
University of North Carolina at Wilmington

APPLIED SOCIOLOGY
TERMS, TOPICS, TOOLS, AND TASKS

Stephen F. Steele

Anne Arundel Community College

Jammie Price

University of North Carolina at Wilmington

THOMSON

WADSWORTH

Australia • Canada • Mexico • Singapore • Spain • United Kingdom • United States

Editor in Chief, Social Sciences: Eve Howard
Executive Editor: Sabra Horne
Assistant Editor: Stephanie Monzon
Editorial Assistant: Melissa Walter
Technology Project Manager: Dee Dee Zobian
Marketing Manager: Matthew Wright

Marketing Assistant: Michael Silverstein
Advertising Project Manager: Linda Yip
Ancillary Coordinator: Candace Chen
Print/Media Buyer: Doreen Suruki
Cover Designer: Bill Stanton
Printer: Webcom Limited

For more information about our products,
contact us at:
Thomson Learning Academic Resource Center
1-800-423-0563

For permission to use material from this text,
contact us by:
Phone: 1-800-730-2214
Fax: 1-800-731-2215
Web: http://www.thomsonrights.com

Asia
Thomson Learning
5 Shenton Way #01-01
UIC Building
Singapore 068808

Australia/ New Zealand
Thomson Learning
102 Dodds Street
South Street
Southbank, Victoria 3006
Australia

Canada
Nelson
1120 Birchmount Road
Toronto, Ontario M1K 5G4
Canada

Europe/Middle East/South Africa
Thomson Learning
High Holborn House
50/51 Bedford Row
London WC1R 4LR
United Kingdom

Latin America
Thomson Learning
Seneca, 53
Colonia Polanco
11560 Mexico D.F.
Mexico

Spain/ Portugal
Paraninfo
Calle/Magallanes, 25
28015 Madrid, Spain

Table of Contents

Preface

This book provides a brief survey of a variety of sociological topics, with emphasis on the practical value of sociology in the 21st century. We wrote this book because we believe sociology can provide much to improve our society's quality of life. Society is built by and is very much in the hands of its occupants. Our goal is that by reading this book you gain new skills and that you use these skills to address a variety of issues in your professional, academic, and personal life. We hope this is true whether this is the only course in sociology that you take or one of many in a career of sociology.

We designed this book for use as an ancillary to virtually any Introduction to Sociology text, by any publisher. We have kept it short and, with the guidance of our colleagues at Wadsworth Publishing, we have structured it to integrate with the most frequent topics addressed in Introduction to Sociology courses. The book will stand alone in introductory classes that draw on multiple learning resources such as the Internet, monographs, and selected published articles. But the book need not stop at the introductory level. It is versatile and can be used in applied sociology courses, social problems courses and action-learning courses at all levels.

The organization of this book takes into account the amount of time students have to study. The chapters are short, formatted for easy reading, and tailored to the time available. Modern learners get their information from a variety of sources. Only one of these sources is reading, more specifically, reading books. In addition, the complexity of life often makes it difficult to set aside long periods of time to study. You often only have a little time "here and there." That which may follow is an "all-or-nothing" perspective. In other words, "I don't have time to study." Hence, "I can't read all the material, so I can't study at all!"

The outcome of this process is a downward spiral. No understanding of the material, reduced confidence in the learning environment (whether online or in a classroom), and less interaction with the learning experience. All of which lead to less learning. We have structured this book so you can "get in the learning game" even if you only have a little time. Think of the format for this book in terms of the amount of time you have. The more time you have to study, the deeper you can go into the material. Here's what we have in mind:

Time Available	Learning Target	Learning Objective
1-5 minutes	Quick Start	Quick read
5-10 minutes	Quick Start, Terms	Quick read, reflect
10-20 minutes	Quick Start, Terms, Topics	Read, reflect
20-30 minutes	Quick Start, Terms, Topics Tools	Read, reflect
30+ minutes	Quick Start, Terms, Topics, Tools, Tasks	Read, reflect, complete tasks

Finally, we would like to thank our reviewers, as well as our colleagues at the *Society for Applied Sociology*, for their helpful feedback and support. Our children and grandchildren provided great motivation during the writing process. They represent to us all children who must live in a complex and increasingly difficult world. As sociologists, we owe them the best possible world that we can build.

Chapter 1. How Can I Use Sociology?

QUICK START

In this chapter you will learn:

- What you can do with sociology.
- The differences among between basic, applied and clinical sociology.
- How to apply the sociological perspective.
- What theories are and their value.
- How to create a theory.
- To use a fishbone diagram to map your theory.

TERMS

Applied Sociology	Using sociological tools to address a specific social problem, often for a particular group of people.
Basic Sociology	Sociology that focuses on directly testing or developing theory.
Clinical Sociology	Using sociology primarily for the purpose of diagnosing and measuring interventions to improve human interaction.
Sociological Perspective	Identifying patterns in human interaction, how and why these patterns exist, the consequences of them, and how to reproduce or change the patterns.
Sociological Theory	An explanation of group attitudes, beliefs, values, and behavior.

TOPICS

You're studying sociology, and before too long you'll ask yourself a common question: "What can I do with this stuff?" It's a good question, and one with a good answer. Let's look at how sociology can be used.

All sciences have at least two thrusts: basic (some people like to call this "pure") and applied. In basic science, the goal is to produce good theory. You are probably asking, "Theory? Who cares?" Before you ignore this important tool, think about this: If you want to fix something, wouldn't it help to know how it works? Theories are efforts to explain how things work. For example, fixing a car's engine would be much easier if we knew how it worked. The better the understanding of how it works, the better and faster we are in fixing it. But theory isn't enough. Having an idea about how a car's engine works and using that idea to actually fix the car are often two different things. So, sciences have applied aspects, too. In this example, fixing the car is the application of the theory. In applied science, the goal is to solve real-world problems. To solve real-world problems, applied sociologists must know theory and how to apply it.

You may be asking: "What do you mean by **applied sociology**?" Well, applied sociology is:

"Any use (often client-centered) of the sociological perspective and/or its tools in the understanding of, intervention in, and/or enhancement of human social life." [1]

Let's dissect this definition. Notice that the definition contains both elements of theory and practice. "Any use" or application suggests knowledge and a distinct approach. The notion of "client-centered" work requires some thought because it really separates basic from applied work. In client-centered work, someone other than the person investigating it presents the problem. These problems can be "dirty" in that they exist in natural settings that may, in the best of times, be less than perfect for collecting information or using theory. Yet they are of great value largely because they are real and need solutions!

Sociological perspective refers to identifying patterns in human interaction, how and why these patterns exist, the consequences of them, and how to reproduce or change the patterns. You will learn more about the sociological perspective in the chapters to follow. For now, understand that there are many ways to "look at" a situation. The sociological perspective is a distinct and powerful way to look at human interactions.

"Understanding of, intervention in and/or "enhancement of human social life" mean that applied sociologists work to advance our collective knowledge of social phenomena, solve problems (intervention), or improve social interaction. Often, applied sociologists work on all three dimensions simultaneously.

We will use the term *applied sociology* throughout this book. We wish it were that simple. Modern sociologists use a variety of terms when they talk about actually applying their discipline. Some sociologists who use the discipline for the purpose of diagnosing and measuring intervention call what they are doing "**clinical sociology**." Others suggest that *sociological practice* is a better term, because it is a term that denotes both clinical and applied. Knowing these different terms for applied sociology should help you when you read other sociological material. For now, let's just stick to "applied sociology."

Here's an example of basic and applied sociology. What happens when organizations change? Much scientific energy has been devoted to determining how organizations work. How big should they be? What should they look like, or how should they be structured? What kinds of leaders are best? These questions are all solid sociological questions for which basic researchers have toiled to answer. Basic sociologists initiate projects primarily because of their own interest in the topic. Whatever the topic, they primarily seek to build or test **sociological theory**.

Now, let's look at an applied angle on this same problem. Suppose someone decides to change the organization in which you work. This becomes a real-world problem, not only for the person who needs to lead the change, but also for the person who lives and works in the organization. Wouldn't it be nice if the people responsible for these changes had a little basic theory under

---

[1] From Steele, S. and Iutcovich, J. (Eds.) 1997. *Directions in Applied Sociology*. Society for Applied Sociology: Arnold, MD. Page 154.

their managerial belts? If this sounds as if theory and practice, or basic and applied sciences, are linked, you are right! Good practice demands good theory. Whether formally trained or not, a good plumber knows a great deal of physics' theory. And a good manager needs a good deal of organizational theory.

In sum, both applied and basic science are needed in any discipline. **Basic sociology** focuses on testing or developing theory directly. Applied sociology uses theory as a tool to solve a client-centered problem. In basic sociology, theory is often the end. In applied sociology, theory is a means to an end. Applied sociology is not atheoretical; rather, theory plays a different role in applied sociology than it does in basic sociology.

In fact, one thing about theories -- everybody has one! In sociology we have three main theories: functionalism, conflict theory, and interactionism. To help us understand and solve a problem, we tend to use one of the three, or a combination of them. All of these theories are useful and powerful tools. However, sometimes none of these theoretical tools work for a given problem. Sometimes none of them, or combinations of them, provides a valid explanation of what is going on or ideas about solutions to the problem. In this case, we have to create our own theory. That sounds daunting, but it is not. Remember, theory is nothing other than an explanation. Applied sociologists, working to solve client-driven problems, often find that they need to create theory to fit the circumstances of the problem.

To create theory, you need some conceptual tools. Some theories are deductive, others inductive. Deductive theory makes sense of situations by taking explanations provided by a theory external to the problem. Deductive theories exist prior and external to the problems that we apply them to. In contrast, inductive theory is created from observing a specific situation or problem. It is problem, or experience, based.

You probably already have experience with deductive and inductive theories in your own lives. Have you ever used your family members or friends' comments to make sense of some experience? For example, let's say you applied for a job, and you didn't get it. Your family and friends tell you, "That job isn't right for you." Or "Maybe you need to gain some more skills and education for that job, or try getting another job in the company and work your way up." These are deductive theories. Sometimes deductive theories work well. Perhaps the deductive theories offered by your friends and family worked for your situation.

Then again, perhaps the deductive theories didn't work. You might have said, "I should have gotten that job. I have more than adequate training for that job." Maybe you do some digging and find out that they hired the boss's son. You suggest an alternative theory, based on your observation. "I never had a shot at that job. They already had someone for the position, and just let me apply to follow procedure." This would be an inductive theory. Sociologists do the same thing. If no prevailing theory about some social situation exists, or existing theories do not explain the situation well, sociologists might begin to make some observations, and create theory from the observations,.

Creating problem-based theory was named "grounded theory" by sociologists Glaser and Strauss. Glaser and Strauss argued that grounded theoretical explanations for why things happen were literally "grounded" in the human interactions and surrounding situation. As such, inductive, or grounded, theory is a very real and powerful way for applied sociologists to understand social reality. Hence, it is very important for applied sociologists to know how to create new sociological theory, as well as to apply existing theory.

TOOLS

Clients often hire applied sociologists to design new programs or evaluate existing programs. These challenges will draw on your ability to use the sociological perspective, to clearly describe the parts of the program, and to think creatively. Whether its a marketing plan, a healthcare program, a crime prevention project, or a program to improve family life, clients are very likely to have their own theories as to how to plan a program, or whether a program works. Clients' theories are based on lived experiences. Applied sociologists need to draw on their clients' grounded explanations to help design a program or evaluate one. To do this, applied sociologists must be able to communicate effectively and listen to what clients say so they can fully understand the situations in which they work.

Here are two tools that can help you understand and use sociology: (1) the sociological perspective and (2) creating theories and using a fishbone diagram.

1. The Sociological Perspective. When it comes right down to it, what special skills and abilities will sociology provide? Let's look at three.

Tool	What is it? How do I do it?
Sociological perspective	The way sociologists view a problem can be a powerful tool in understanding social issues and how people act. We want you to try the components of this perspective and continue to practice them hereafter.
Looking at the "big picture."	Sociology compels people to focus on the social situation -- the big picture. It supports this view whether looking between two individuals, among groups, or among societies. Some job applications of this skill include organizational design and development, project design and management, continuous improvement, strategic and community planning, policy formation, and forecasting. 1. Start by looking at social interactions at the next level up from the person and personalities. For example, can you see personal problems in getting an education from the point of view of a student or instructor? 2. Then, work your way "up" by moving to the next level. In this example, consider your class, your school, education in our society, global education. 3. Now consider the forces that influence each of these levels. What impact are these influences and levels having on the "person" involved in the situation?

2. Creating theories and using a fishbone diagram. One useful way to apply theory is to create a fishbone diagram. A fishbone diagram is another skill that you can use in business, government, and other venues to make sense of a situation and problem solve.

Tool	What is it? How do I do it?
Fishbone diagram	Sometimes you need to create your own "road maps." While functionalism, conflict theory, and interactionism will give us some direction, we may need to "start from scratch" in building a model of just what seems to be happening.
Getting started	1. Identify a social problem or condition that you need to understand. This is the "effect or the outcome." 2. Use brainstorming or observations to identify the main causes of the problem. Group these into meaningful categories if necessary. Often, the main categories of causes are resources, methods, people, materials, etc.
Using the fishbone diagram to map your theory[2]	1. Take a piece of paper (notebook paper, paper from a flip chart, whatever size is relevant for the project). Turn it on its side ("landscape style" in "computerese"). You'll need a pencil, pen, or marker depending on the size of the paper and the audience. Incidentally, you can do this on a computer. Many of the common graphics programs can help here. 2. Now start your fishbone diagram. Draw an arrow from left to right. At the right end of arrow write "effect." Above the this line, write the word "Causes." Now, you have something that looks like a fish's backbone. Something like this: Causes ⟶ Effect Causes 3. Now write the names of the causal categories that you identified in step 2. Write some above and below the line. Then "connect" the specific causes to the categories. Draw an arrow from each cause to the "backbone," like this:

[2] Adapted from Brassard, M. and Ritter, D. (Eds.) 1994. *The Memory Jogger II*. GOAL/QPC: Methuen, Massachusetts. Pages 23-32.

	Causes Category 2 Category 3 Ex: Demographic Cause Age, Cause Gender, **Effect** Ex: Change in Crime in Our Area, Cause Income, Cause Education, Category 1 Ex: Socio-economic, Category 4 **Causes** 4. Finally, interpret your results by looking for "root" causes. To do this, ask, "Why does it happen?" for each of the causes written in step 4.

Express it in words	Now that you have a map of the presumed "causes and effects":
	1. Write a brief summary of what you've found. Consider starting out by presenting the problem or situation, such as "The client reported that..."
	2. After brainstorming and reviewing your fishbone diagram, write: " x (the number of presumed causes) factors influence the client's problem. These include:
	    a.
	    b. etc.
	3. Write a summary statement.
	4. Present this to your client.

Think about what happens next	1. For some clients, just discussing how things work may be enough. But with others you may need to think about how you plan to measure your causes and effects. 2. Reflect on the social theories you've learned. Can functionalism, conflict theory, and interactionism help you make more sense of just what's happening here?

TASKS

Some of you reading this may not become sociology majors, but you'll be stronger, no matter what you do, if you add a sociological approach to your personal tool kit. The views and perspectives that you are learning in sociology can only be improved by applying them. Here are some exercises to help you apply sociology.

1. A city is running out of room in its landfill (i.e., garbage dump). City leaders are looking for a town in the surrounding area willing to sell its land to serve as a new landfill. Use your sociological perspective to identify the pros and cons for a town considering placing a bid for the landfill. The landfill will bring jobs and revenue to the town, but it will also bring a host of possible problems, namely, environmental dilemmas. Use your sociological perspective to identify the pros and cons to the town. What social factors shape the pros and cons you identify? How will the pros and cons you identify shape the town's culture and society?

2. Use your theory creating ability to tackle a real-world problem. Select a client (this could be a friend, a neighbor, your boss, another student).
   - Interview this client to determine the nature of a social situation or perceived problem that is of importance to this person or organization.
   - Then employ the theory construction processes and the fishbone diagram to make sense of this situation. To add value to the process after you have finished the theory, brainstorm some recommended solutions.

Chapter 2. Model, Measure, and Make Sense

QUICK START

In this chapter, you will learn,

- Why theory is an important, practical tool.
- To demonstrate sociological imagination and creativity in solving problems by using theories.
- How to measure a social variable, and at what level of measurement.
- How to perform the basics of quantitative and qualitative data analysis.

TERMS

Code	A symbol scheme, usually numbers, that provides a shorthand way of representing the values of a social variable.
Conflict Theory	The explanation of social interaction that, in short, states that society is the result and cause of conflict. Contrasting values and desires to possess social resources explains social behavior.
Constants	Measures to which responses don't vary. All responses are the same.
Closed-Ended Question	A question that has a defined set of mutually exclusive responses.
Data Analysis	Searching for patterns in the information collected so you can make sense out of social reality.
Data Collection Method	Ways to collect data on individual and group characteristics, attitudes, beliefs, values, and behavior, such as surveys, interviews, observation, and experiments.
Functionalism	The explanation of social interaction that, in short, states that society is a system that strives for balance. The interdependent parts of the system, such as institutions or roles, exist because they provide a function.
Frequency	The number of responses for each value of a variable in a sample.
Frequency Distribution	A table displaying the frequency and percentage of all of a variable's values.
Index	Using multiple indicators to measure abstract or complicated concepts, such as quality of life. The indicators and their responses are grouped, often summed, together to create a more valid measure.
Indicator	A question used to measure a concept that can't be directly observed or measured.
Interactionism	The explanation of social interaction that, in short, states that people create and change society through social interaction.
Interview Guide	A list of the topics or questions that a researcher wants to raise in an in-depth interview.
Likert Scale	A response set that reflects a continuum of agreement with a statement about the social world. For example, "How satisfied are you with your college education? Strongly Dissatisfied, Dissatisfied, Satisfied, Strongly Satisfied."
Measure	Ways to observe or indicate a person's or groups' values, attitudes, beliefs, and behaviors.

Measurement Instrument	A collection of measures used in a study, such as a survey questionnaire or interview guide.
Mutually Exclusive	When the values of a measure do not overlap.
Observation Guide	A list of the types of individual behavior or group interactions that a researcher wants to observe in an observation study.
Open-Ended Question	A question without a defined set of responses for a respondent to choose from.
Qualitative Measurement	Ways of measuring that systematically observes attitudes, beliefs, values, and behaviors in their entirety.
Quantitative Measurement	Ways of measuring that breaks down attitudes, beliefs, values, and behaviors into their component parts and marks or codes the component parts with numbers.
Variable	A measure of a concept, the value of which can change across people or cases.

TOPICS

This chapter is longer than the rest. We did this because it explains three important and somewhat complicated things that are best connected. To apply sociology is to create a representation of how things work (a model). The model isn't much good unless we can demonstrate that it works or doesn't (measure). Finally, our work is useless unless it is communicated effectively (makes sense). So, this chapter takes you step by step: model, measure, and make sense.

You don't need us to tell you that societies are complicated things. Sociologists try to make sense of society. Like other sciences, sociology builds theories to help explain how society operates. Theory often gets a bad rap. People often think of theory as, at best, boring, and, at worse, useless. What we would like to do in this chapter is change your mind about the practical value of theory. We think that, even if you never take another class in sociology, sociological theory can empower your life. Theories explain life and help us make sense of what's going on. As such, theories are tools, providing frameworks for understanding and change.

Modeling

Social theories provide "road maps" that help us understand how society works. Simply speaking, how are we going to fix societal problems if we don't know how society works? When we look at the complexity of society we look at millions and perhaps billions of individual people. Yet, despite these large numbers, most people behave in patterned ways. They act more similarly than they do differently. Why and how does this happen? Sociology addresses this question, with an array of theories, that many sociologists categorize into three general theoretical perspectives: **functionalism**, **conflict theory**, and **interactionism**. Most sociologists blend these three perspectives when trying to explain social life. Here we provide a quick review of each perspective's focus and concepts.

11

Let's take a common problem during the holidays: overbooked flights. We can use our three sociological theories to explain why this problem occurs. Functionalism would see a gap between the air travel system and traveler needs. An obvious solution may seem to add more flights. However, increasing flights is not easy, even in a good economy. It requires more airports, with more space to land, and more resources to control increased air traffic. Airports are relatively fixed systems. It is difficult and costly to increase the number of airports. So when airlines respond to customer demand by scheduling more flights, the system becomes clogged, creating more delays. Given this context, what can airlines do? Perhaps it is time to change passenger needs by providing more services in airports for bumped passengers such as workstations, Internet access, meal vouchers, mini-massages, or movies to watch.

Rather than a gap between system and needs, we might suspect that the airline companies are taking advantage of passengers whom they know have few alternatives to holiday air travel. Seeing the problem in this way would be adopting a conflict perspective. Conflict theory sees problems as a clash between people or groups with power to exploit or oppress others. When an airline announces that a flight is overbooked, passengers could, theoretically, forgo their seat on that plane and buy a ticket for another flight. This option only really exists in larger cities where more than one or two airlines service the airport. Even in large cities, most passengers could or would not purchase a new ticket because airlines dramatically increase the cost of tickets purchased within two weeks of the flight. Even this option may not work as most airlines overbook flights. Given this context, what can airlines do? Conflict theory would push us to instead ask what consumers can do. They could organize their efforts and present a unified case to the airline executives not to overbook flights. Collectively, consumers have much more power than do individual consumers to affect change. They could boycott air travel and picket outside airports across the country to bring more attention to the problem.

Finally, we could see the problem of overbooked flights during the holidays in the eyes of an interactionist. Interactionism looks at how actors define situations and at what influences these definitions. In this case, most consumers define the situation as a problem created by the airlines. What influences this belief? They may have heard about this problem in the news, and from friends and family. What information is presented, and how it is presented, influences how people define situations. For example, if overbooked flights receive a great deal of media coverage, and if most of the coverage highlights passenger suffering, then consumers may grow increasingly discontent. They themselves may have experienced no serious hardship due to overbooked flights, but they may take the role of other, and identify with those that do. Given this context, what can airlines and consumers do? Interactionism would encourage the two groups to discuss the problems together, and to try and understand each other's perspective. Then, from this understanding, owners and consumers could construct a mutually agreed upon solution.

Measuring

Now, let's shift our attention to what you need to do after you have a model for how things work. One of the key things that we have learned in science is that we need to be able to measure something in order to study it. This is true whether your doing biology, chemistry, psychology, economics, or sociology. Of course applied sociology is no different. If you can't measure a problem, then you can't determine whether the interventions or programs you built to solve the problem worked! Hence, **measurement** is critical to applying sociology.

Let's look at an example. Suppose a travel company hires you to determine how satisfied people are with their vacations. The company wants to know this so that it can recommend enjoyable vacations to all of its clients. Seems easy. But how are you going to determine people's satisfaction? That is measurement. We need to develop a way to measure people's satisfaction with their vacations. We could ask them, "Would you recommend this vacation to your friends and family? Yes or No." Or we might ask, "How satisfied were you with your vacation? Are you strongly satisfied, somewhat satisfied, somewhat dissatisfied, or strongly dissatisfied?" Or we might go into the company's files and see what vacations people take repeatedly. Taking the same vacation more than once surely serves as an **indicator** of satisfaction.

Additionally, sociologists often want to transfer or **code** the answers to questions into numbers. For example, the responses "strongly dissatisfied" might be coded 1, "somewhat dissatisfied" might be coded 2, "somewhat satisfied" might be coded 3, and "strongly satisfied" might be coded 4. These numbers then become the data, not the written words. Sociologists use the numbered data to do statistical analyses. This idea of turning observations of social reality into numbers may be new to you, but it is basically the same thing as turning observations of physical reality into numbers, which you have been doing since you could talk. How many times have you answered the question, "How old are you?" That number is nothing other than a way to measure physical reality or the passing of time. People created that measure and gave it meaning over time. Same goes for the answers to "How tall are you?," "How much do you weigh?," "How much does it cost?," "What time is it?," and so on.

Not unlike the physical world, we can measure the social world. We call the tools that we create to measure the social and physical world **measurement instruments**. The physical sciences have more exact measurement instruments, such as thermometers to measure temperature, scales to measure weight, or X-rays to measure cell growth. For the most part, they are not measuring things as slippery as attitudes, values, and beliefs. Nor do they have to rely on their research subjects to verbalize the information they are trying to measure. For example, a physician may ask you how you are feeling, but then she will also likely run a battery of lab tests to measure that same thing. It is the latter she will rely on to diagnose and treat you. Sociologists do not have that luxury. We must design measures that accurately reflect the concept at hand and questions that people can and will answer.

Finally, before we try to actually create measures, we need to consider one more issue. The measurement issues we've discussed above largely pertain to **quantitative measurement**. Applied sociologists also use **qualitative measurement**. We've explained the difference between quantitative and qualitative approaches earlier in the book. Here we just want to remind you that qualitative and quantitative approaches to measurement differ. Qualitative measurement jumps directly into the social world, allowing the researcher to be the measurement instrument through his or her interactions in and observations of the social interaction. In this sense, qualitative measurement bypasses the need to create a standardized measurement instrument with questions and responses that will provide an accurate window on the social world. Instead, with qualitative measurement, researchers create an **interview or observation guide**.

An interview guide lists the topics or questions that you want to raise when you talk with people in in-depth interviews. Similarly, an observation guide lists the types of individual behavior or group interactions that you want to observe. Each guide is supposed to do just that -- guide, not dictate, the questions asked or observations sought. When using qualitative measurement, the researcher has the flexibility to ask other questions that may arise from the discussion, or look for other interactions as suggested by whatever he or she observes. Let's look at an example of what you would put on interview and observation guides.

Suppose a local child day care center wants to improve how children interact with other children in their center. They have noticed that when new children enter a classroom, the children jockey for attention from each other and the teacher. You decide you will need to both observe the children's interaction, and then interview some of the children who interact well when new children join the class. On your observation guide, you create sections for the different group characteristics, attitudes, values, beliefs, and behaviors that are salient to this project. Two might be group cohesiveness and boundary maintenance. Underneath each section you list some specific behaviors and interactions that you might observe that would indicate each respective group characteristic. For example, underneath group cohesiveness you might list "children sharing/not sharing toys" or "nicknames." Underneath boundary maintenance you might list "inviting/rejecting others to join a game" or "existence of sub-groups or cliques." You get the idea. Then while you observe the children's interaction or soon afterward, you write down examples of whatever is on your observation guide. You revise the guide as you go, as is indicated by your observations.

An interview guide would follow the same process. Instead of listing behaviors or interactions, you list questions that you would like to ask the children, under their respective headings. For example, you might have a section on family life, the day care center, and new classmates. Then under each section you list specific topics or questions. Under family life, you might list "number of siblings," "position in sibling order," "parent's marital status and relations," and "a regular day at home." Discussion of these topics would indicate the quality of the child's home life, and his or her experience handling conflict or change. You would have similar kinds of topics and questions listed under the other sections. In observing all the children and interviewing the children who consistently play well, you might find that children without siblings or children who are the youngest in their families are primarily the children expressing difficulty when new children join the class. These children might not have the same skills in handling change as children who have younger siblings. When a new child joins the class, they perceive a state of normlessness, and struggle to create a new social order. During this struggle, they act out. One of your recommendations might be that the day care center should routinely play games where the children rotate onto different teams so that they learn to deal with change and work with others.

Making Sense

When it comes right down to it, we can model and measure all we want, but if we don't make sense of the information our work is lost! In this section, we provide you with a quick introduction to the basic **data analysis** techniques. We think the basics below will help you both as citizens and as professionals. In general, applied sociologists use two approaches to analyze data that they collect on a problem: quantitative data analysis and qualitative data analysis. Let's look at each of these approaches.

Quantitative Analysis	Qualitative Analysis
Look for patterns in numbers through the use of statistics such as percentages.	Look for patterns in written text, notes, documents, audio recordings, or video recordings.

As an applied sociologist, whether you use quantitative or qualitative research methods in trying to understand and solve your client's problem, you will likely gather a boatload of information, otherwise known as data. How will you make sense of this information? How will you learn from the data you collect? The answer is data analysis. In this chapter we will introduce you to the world of data analysis. These data analysis tools will serve you in any social or physical science pursuit.

The first step in data analysis, regardless of **data collection method**, is to answer the question, "What's going on here?" This is basically a descriptive question. It asks you to describe the initial facts about the problem at hand. An answer would include the number of people included in your study, their characteristics such as demographic information, and a description of the setting (if applicable). To get started in answering this question, you should calculate and display the frequency of each value for all of your **variables**. A **frequency** refers to the number of times something happens. Central to describing something is to count the number of times that something happens. A **frequency distribution** contains both the frequency of each of a variable's values, and each frequency expressed as a percentage. Suppose we have asked a random sample of Americans their opinion on whether the U.S. military should accept homosexuals. Let's compute a frequency distribution on this variable.

Compute a Frequency Distribution
1. Identify the target group you want to describe. Is it all the participants or respondents in your study? Or a sub-group, such as only the residents of a particular county, or only the men, or only people of a particular ethnicity? In this case, we want to use all respondents, but later we will look at only male respondents.

2.	Determine the variable you want to describe. Out of one study you may have dozens of variables. Which one(s) is of interest now? How is it measured?
	Should the U.S. military bar homosexuals?
	1. Yes
	2. No
	8. Don't know
	9. No answer

3. Count the number of people who responded with each response category:

Response	Frequency
1. Yes	573
2. No	301
8. Don't know	88
9. No answer	38
Total	1000

4. Translate the number of responses into percentages (also called relative frequencies)

Response	Frequency	Percentage (%)
1. Yes	573	57.3
2. No	301	30.1
8. Don't know	88	8.8
9. No answer	38	3.8
Total	1000	100

5. Use the percentages to calculate a cumulative percentage column. This column adds each percentage together, moving down the column.

Response	Frequency	Percentage (%)	Cumulative Percentage
1. Yes	573	57.3%	57.3
2. No	301	30.1	87.4
8. Don't know	88	8.8	96.2
9. No answer	38	3.8	100
Total	1000	100	

Qualitative Analysis: In qualitative data analysis, you need to systematically and rigorously analyze all of the data collected. To keep this process from being overwhelming, we need to break it down into the following steps.

Steps to Qualitative Analysis
1. Read all of your data (transcribed interviews, interview notes, observation notes), making notes on any possible patterns you identify while reading, as well as any concepts that make sense of those developing patterns.
2. Study the notes and ideas that you wrote while reading your data. Take your knowledge of the topic and the project and try to make sense of the patterns and concepts identified in your notes. Let your knowledge inform your original list of patterns and concepts. Then revise your list, giving each pattern and concept an abbreviated name or code.
3. Refocus your analysis on the patterns and concepts developed at the end of step 2. Read all of your data again, circling, highlighting, or in some way marking off the text illustrating each pattern and concept. Write the code name next to each circled segment of text.
4. Repeat step 3. Here you may code your data exactly as you did in step 3. If so, you have a sense of strong validity and reliability of your analysis. In contrast, you might change the codes on the same segments of text. Or you may code additional segments of text with the existing patterns and codes. You may even identify entirely new patterns and concepts and codes for them in this step. You may want to repeat step 4 again, after a week or two, as another check on the validity and reliability, especially if you make any changes during the first run through step 4.
5. Record how many times each pattern and concept is coded in step 4.
6. Pull out the coded segments of text. Read these again, looking for relationships between them. For example, what patterns or concepts occur in the presence of others? Which usually come before or after others? Which only hold for sub-groups of your entire sample? Which patterns evolve differently for different sub-groups? It may help to develop new codes for these relationships, and then to code each segment and sort the segments accordingly. You may need to repeat this step several times. Make notes of your "findings" as you proceed through this step.
7. Record how many times each relationship between patterns and concepts occurs in step 7.
8. Write in paragraph form a description of the patterns and concepts you found in step 4. Then write in paragraph form a description of the relationships between the patterns and concepts that you found in step 6. This process will take several drafts. Your goal should be to "tell the story" of your findings. Include how often the patterns, concepts, and relationships occur.
9. Choose excerpts from within the segments of coded text to illustrate the patterns, concepts, and relationships. Weave these excerpts into the story you wrote in step 8.

Many sociologists find it helpful to write about what they are finding at each step in the qualitative data analysis process. Writing what you are thinking and learning can clarify and push your reasoning.

TOOLS

Modeling Tools

As complex as human society is, using one or more of these theories can give us a grasp on what is happening and what could happen in our society. Theories are not just vague, passive generalizations -- they help explain social action at all levels. You can use the table below to guide your application of sociological theory to any social problem or issue.

Tool	What is it? How do I do it?
Function-alism	1. Start with this basic assumption: Society is a set of interdependent parts. 2. Choose a social situation. 3. What are the needs of this situation? That is, what is supposed to be done here? 4. Think about this situation as a system. Ask yourself: What are the parts that make up the system? What function does each part serve? What are the parts' relationships with the other parts? 5. Draw a diagram inside the needs circle with arrows that show the connections between the social parts. 6. Are there any "gaps" between what is supposed to happen and what the system is able to do?
Conflict theory	1. Start with this basic assumption: Conflict is frequent and social change is certain. 2. Choose a social situation. 3. What groups have more power to control other people and the outcomes of the situation? List them. 4. What values, interests, and goals does the group have in this situation? 5. What groups have less power and less influence over the outcomes? What values, interests, and goals does the group have in this situation? 6. Does the group with power think it can achieve its goals? Does the group without power think it can achieve its goals?
Interactionist theory	1. Start with this basic assumption: Societies and persons in them are guided by the way things are collectively defined. 2. Select a social situation. 3. What are the different definitions of this situation? What are the characteristics of the people holding these different definitions? 4. What influences their definitions? 5. How does their definition influence how they behave in daily social life? 6. How does it influence how they interact together? 7. How do these definitions influence how people act?

Measuring Tools

You can see that creating measurement instruments is pretty creative stuff! But measurement can get a bit tricky too. So, let's look at some measurement concepts that will help you develop measurement instruments. Remember, though, you must start any project by clearly identifying the problem and related issues. Without that step, measurement tools will not help. Keep in mind that the more complex the individual or group characteristic, attitude, value, belief, or behavior you are studying, the more difficult it is to measure.

Tool	What is it? How do I do it?
Group vs. individual characteristics	In general, group characteristics are harder to measure than individual characteristics. Take, for example, group cohesiveness from the example above. Group cohesiveness requires more sophisticated measurement than, say, the individual characteristics of education, race, and religion.
Attitude, value and belief vs. behavioral questions	Behavior is often easier to measure than an attitude, value, or belief. We can look at data previously collected about people's behaviors, such as their criminal record, health care services received, or whether someone took a vacation at the same place twice. Or we can observe a particular behavior happening in real time, such as how children play together. We can ask people about their behavior, such as "Have you ever smoked marijuana?" or "Do you usually vote Democratic, Republican, or for third party candidates? If the question is written well and the issue salient, people can generally recall past behavior.
Variables vs. **constants**	You need to make sure that your measures will have some variation in response across individuals or groups. If not, you have a constant, which may complicate your statistical analyses. Often constants occur because a question did not provide responses that reflect the true variation in the population. For example, you ask fraternity members the number of alcoholic drinks they consumed per week and provide the following responses: 0, 1-2, 3-5, 6 or more. All fraternity members respond "6 or more." Your categories do not reflect the true variation. You need to break 6 or more into multiple categories.
Closed-ended question	A question with a defined set of **mutually exclusive** responses, such as: What is your sex: (1) Female (2) Male Is your best friend a: (1) Woman, (2) Man, (3) Neither, (4) I don't know, (5) I don't have a best friend. What is your current marital status? (1) Married, (2) Separated, (3) Divorced, (4) Widowed, (5) Single.
Open-ended question	A question without a defined set of responses, such as: What was your first job? _____ Sometimes the recorded responses are then organized into quantitative categories, such as different occupational categories.
Index	Some concepts require multiple indicators because they are very complicated or abstract. Using multiple indicators to measure a complicated concept is a form of triangulation, giving us a better view of reality. For example, you can't measure quality of life with one measure. It has several dimensions to it, such as crime rates, pollution, unemployment rate, cost of living, quality of educational institutions, health care services, parks and recreational amenities, weather, etc. If we wanted to measure quality of life, we would create a measure for each of these dimensions. Then, to summarize quality of life, we would sum each respondent's answer to all of the questions, thereby creating an additive index.

Likert scale	A Likert scale is a response set that reflects a continuum of agreement with a statement about the social world. For example, to what extent do you agree or disagree with abortion?
	1. Strongly Disagree
	2. Disagree
	3. Undecided
	4. Agree
	5. Strongly Agree
	Sometimes the agree-disagree labels do not fit the issue at hand. In this case, applied sociologists would use labels such as "Frequently," "Some," "Rarely," "Never," or "Very Unlikely," "Somewhat Unlikely," "Somewhat Likely," "Very Likely." Applied sociologists may omit the middle or neutral category if all respondents should have an opinion on the issue at hand.

Making Sense Tools

Tool	What is it? How do I do it?
Graphics: pie chart	Use a pie chart to graphically summarize data on one variable. Below, we use Word to create a pie chart from data on Americans' attitudes about the environment. To create a pie chart in Word, pull the "Insert" drop-down menu, click "Picture," then "Chart," and then pull the Chart drop down menu and click "pie chart." You can also use many other software packages to create pie charts.
	Humans have the right to modify the natural environment to suit their needs.

Graphics: bar chart	Use a bar chart to summarize data on two variables. Below we use Word to create a bar chart with data on Americans', with and without college degrees, attitudes about the environment. To create a bar chart in Word, go to the "Insert" drop menu, then to "Picture," then "Chart." If a bar chart table does not appear, pull the Chart drop down menu and click "bar chart." You can also use many other software packages.

Humans have an ethical obligation to protect the environment. |
| Graphics: table | Use a table to show raw data on one or two variables. Below we use Word to create a table, using the "Table" pull down menu. You can also use many other software packages.

Do you donate money to an environmental group?

	Under 40	40 or Older
Yes	37.2%	13.4%
No	62.8%	86.6
Total	100%	100%	

Not unlike other tools that we have discussed, communicating is a combination of art and science. You will develop your own style, but you must practice! Here are some basic tools that will apply to most situations.

TASKS

1. Your client is a local public high school with declining SAT scores. Large percentages of children who attend this school are from one-parent homes, and many are children of color. Use each theoretical perspective to explain the decline in SAT scores and to make suggestions on how to improve the scores. Use the table below to guide your thinking.

2. The U.S. Marine Corps hires your company to help develop programs to improve soldiers' marital relationships. The Corps's data suggests that its soldiers have one of the lowest rates of marital satisfaction in the country, and that this is one of the main reasons why soldiers

leave the Marines. You developed a program and implemented it. Now you must determine whether the program you developed is working. To do so, you must determine whether couples are more satisfied with their relationship now.

- How could you measure marital satisfaction?
- How could you measure marital satisfaction in an in-depth interview with each spouse?
- How could you measure marital satisfaction through observing couples interactions?

3. You work for a city manager. A group of city employees is picketing over domestic partnership rights in front of the city hall in a suburb of Burlington, Vermont. The employees want the city to extend health insurance benefits to the romantic partners of gay and lesbian city employees. Recently, the state of Vermont extended marriage rights to gay and lesbian couples. These employees argue that, given that the state now recognizes gay and lesbian marriages, the legal partners of gay and lesbian employees should be treated equally like the legal partners of heterosexual employees. The mayor and the city manager want to determine how much support there is for domestic partnership rights among the city's residents.

- Identify what main data collection and sampling method(s) you will use.
- Determine what topics for which you will need to write questions.
- Develop questions and question stems to measure those topics.
- Identify any potential problems with your measurements.

Chapter 3. Concepts, Culture, Socialization, and Social Structure

QUICK START

In this chapter, you will learn:

- How to use the sociological perspective to solve problems.
- How to observe social interaction and identify norms.
- To recognize where and how you and others use power, prestige, and wealth in social interaction.
- To apply a set of basic sociological concepts to understand our world.
- To define a social setting as something in itself, rather than as the sum of the individuals added up.
- To brainstorm a social situation to determine the social forces that are at work.

TERMS

Concepts	Terms to denote a set of ideas that help us employ the views of sociology in a practical way.
Culturally Relative	The notion that a group's way of life, its meanings, attitudes, values, and behaviors are relative to the context of that culture and can only be understood within that context.
Definition of the Situation	How people in a situation perceive and understand that situation.
Ethnography	A method for observing and writing down the observed patterns of people in a community, culture, or organization.
Ethnocentrism	The belief that one's own culture is better than another.
Internalize	The point at which your group's culture becomes a part of your own individual beliefs, meanings, values, attitudes, and behavioral expectations.
Norms	Rules for social action. These can be formal (i.e. laws), informal (folkways), and/or sacred (mores).
Role	Roles are sets of behaviors, expectations, and obligations associated with a particular social position.
Situation	The social setting in which an interaction occurs.
Social Controls	Ways of keeping people in line with the norms of a group, society, or culture by using informal or formal sanctions.
Social Forces	Collective definitions and social conditions that act on us. For example, the number of people at a particular age that are alive in a society will have an impact on how that society organizes itself.
Social Institutions	Sets of values, norms, beliefs, attitudes, behaviors, and expectations surrounding important aspects of life, such as marriage, family, religion, education, politics, economics, health, science, military, etc.
Socialization	The process through which people learn cultural and societal expectations for human interaction.
Social Stratification	Ways that groups of people distinguish one another and distribute social resources accordingly such as social class, gender, race, ethnicity, sexuality, age, and disability.
Social Structure	The combined influence of all the social institutions and social stratification on individual's and groups' behavior.
Sub-Culture	A way of life that deviates from the larger culture on some main cultural theme.

TOPICS

Similar to other sciences, sociology uses **concepts**. Sociological concepts are sets of ideas that make sense of everyday life. Sociology has many concepts; you have learned some of them in the previous chapters. Now we can just imagine what you are thinking. "I've got to learn a bunch of new words just so I can make sense out of everyday life?" The answer is yes. Here's why: Concepts serve at least three important purposes. First, as stated above, they provide a framework on which to describe social reality. They give us a way of knowing about things. Second, they provide a shorthand way of expressing a set of ideas. Third, concepts serve as blocks to build sociological theory, which explain social realities.

Take the concept of **role**. Roles are sets of behaviors, expectations, and obligations associated with a particular social position. Groups of people over time develop roles to ease social interaction. Roles help people know how to behave and how to expect others to behave. For example, we expect police officers to value and uphold the law, and police officers know they are obligated to follow a set of rules regulating their authority over citizens. Other behaviors in the role of police officer include wearing a uniform, carrying a gun, observing people's behavior, and arresting law breakers. It is easier to use the term *role* to denote all of the behaviors police officers enact than it is to list them all every time. In sum, concepts help us categorize reality in a simple way.

The concept of **situation** is a powerful tool for sociologists. By *situation* we mean the social setting in which an interaction occurs. Let's take a look at what we mean. Suppose two people are walking down a sidewalk. They approach and pass one another, and go on their individual ways. Simple enough! But what is the sociological perspective on this, and how can sociological concepts help us understand this interaction? What sorts of things influence this interaction?

- What are the genders, ages, and races of the two people walking down the sidewalk?
- What are the roles of the two people walking down the sidewalk?
- Where is the sidewalk?
- How wide is the sidewalk?
- Is there anyone else around?
- What are the social classes, genders, ages, and races of the other people in the setting?
- What are the rituals for walking on sidewalks in this subculture?
- How do the two sidewalkers **define the situation**?
- How do the other people in the setting define the situation?

Notice from each of these items that we ask questions about the setting. Sociologists are more interested in the interaction between the individuals in a situation than they are the individuals. Individual personalities, even body chemistry, might affect how the actors act in the sidewalk. Sociologists argue that **social forces** are usually even stronger influences on human behavior than psychological or biological characteristics. All of the items listed above point to different social forces. A person walking down the sidewalk would likely act very differently on that sidewalk if there was no one else around than if other people were there. This might be especially true if the sidewalk was in a busy city. Now let's take this view to another level. We introduced you to the sociological perspective in a previous chapter. Now we are going to focus exclusively on it and its applications. Just what exactly does it mean to be sociological? As we said before, sociologists see most human behavior as social. That is, they think most of how we act with ourselves and with other people is due to the influence of other people.

The nature versus nurture debate may help illustrate this point further. If you had to explain why more women invest more time in domestic labor (cooking, cleaning, family care) than do men, what would you say? If you say that it is because women have some biological or psychological predisposition for either wanting to do these activities or doing them better than men, then you would be emphasizing the influence of nature on human interaction. In contrast, if you said that women do more domestic labor because they were taught to, or because that is what women and men expect women to do, then you would be emphasizing the influence of nurture. Sociologists point to "nurture" as the explanation for most of our behavior, attitudes, and values. In this case, nurture is another way of saying the influence of other people.

The influences of other people can be grouped into two large categories: cultural influences and societal or structural influences. Let's look at culture first. Culture refers to a group's way of life, its shared meanings, values, beliefs, and behavioral expectations. For example, in the U.S., we have a shared meaning for our national flag and a set of expectations for how to physically handle the flag and show deference to it. We have shared meanings and expectations for when and how to tip service workers. These shared meanings for tipping vary widely from those in other countries. People create culture through interacting with one another over a long period of time. We **internalize** culture through interacting with others in our group(s). The process of learning through interaction with others is called **socialization**. An understanding of culture is central to understanding the people who live in that culture and their actions.

Now let's turn to society. Society refers to groups of people who interact often, usually in some common space, and the rules they develop to help manage and regulate these interactions. Many sociologists use the term **social structure** to specifically denote the rules or **norms** of managing and regulating social interaction. In order to ease their repeated social interactions, people develop rules of interaction and **social controls** to enforce them. Imagine if every time you had to interact with another person, you had to negotiate the rules to that interaction. This is what dating is all about and it is why dating can be awkward and uncomfortable. For most of our other interactions, we have a good idea of how we are supposed to act and how others will interact with us. For example, you know how sellers and buyers interact, you know how students and teachers interact, and you know how people interact in religious settings.

With an increased division of labor, people become more dependent on one another, and the rules of interaction become more important. Think of all the work you depend on others to provide every day: people to grow food and others to transport it to grocery stores, people to deliver your mail, people to manage your savings, and coworkers to do their part of their jobs. A sports team symbolizes the dependency of all of us on others. All the team members must depend on each other to do their part. If someone fails, the whole team fails. Like on sports teams, people are paid to do many of the actions we depend on them for, but other actions come with no monetary compensation. We perform them based on trust and commitment to the group alone. For example, we depend on other people to parent their children. We depend on people to respect our homes and other possessions. We depend on others to follow the rules of the road when driving. If people do not follow the rules, we could not interact safely or effectively, and our group would disband. Some of these rules are written down in laws; others are transmitted through culture. Some of the rules pertain to all settings, but most are specific to certain settings.

Sociologists call sets of rules and social controls for specific sets of common behaviors **social institutions**. For example, we have the institution of marriage and family to regulate romantic and parenting relationships. We have the institution of education to manage how we formally teach basic information for interaction and survival in our society. How do institutions come into existence? Over time people develop ways to perform necessary functions, such as how to make money or trade (the institution of the economy), how to explain the unknown (religion), how to discover new knowledge (science), how to organize power and authority to make decisions for large groups (politics), how to formally control behavior (law), how to provide or obtain health care (health), and how to physically protect large groups (military). A social institution represents a set of social behaviors that are necessary for a society to function smoothly. People need these institutions to provide stability and consistency, so that we can interact easily with one another and be more productive. Sociologists focus on how institutions come about, how they affect people's behavior, how people change institutions, and how institutions reinforce inequality. We will discuss these institutions throughout the rest of this book.

The term *society* also includes rules about the social resources of power, prestige, and wealth. The result of these rules is **social stratification**, or divisions between people who have social resources and those who do not. Stratification serves several purposes, of which we highlight a few. For one, it protects the resources of those who possess them. People with social resources develop and enforce rules to protect their resources. Hence, stratification reproduces inequality or the status quo. Stratification also can ease social interactions. Because of stratification, when you enter any interaction you can quickly scan those in the setting and have a good idea of how you and they should and will act. For example, when you are a patient seeing a doctor, you and the doctor both know that he or she has more power and prestige in that setting, and that guides how you interact with one another. You wait for the doctor to see you; he or she does not wait to see you! Similarly, when a police officer asks to speak with you, most people obey.

Some stratification seems legitimate to some people in certain social settings. For example, most bosses and employees expect bosses to have more power, prestige, and wealth than employees. However, most stratification is not legitimate or equitable. For example, in our past and present, many Americans have also expected people with light colored skin to have more power, prestige, and wealth than those with darker colored skin. Many Americans have taken steps to prevent people who are different from them from gaining social resources, resulting in classism, racism, sexism and heterosexism, among others. So, while our social structure eases repeated interactions with other people, it also creates and reproduces inequality.

26

TOOLS

We can use the concepts of culture and society as valuable tools. Let's take an example. A large corporation hires you to improve how the sales department works together. The owners point to low morale in this department, high numbers of sick hours, and high numbers of employees starting workdays late. In order to understand why and how they interact as they do, you will need to look at the culture and society within the group of coworkers and among the entire organization. Any modern corporation can be seen as a complex corporate culture and society. For those working there, the corporation becomes a way of life. This is evident every time you hear, "That's the way we've always done it" in response to a question about why they do their job the way they do. In other words, they're telling you that they follow the norms of the organization. They're conforming to what they believed to be the corporation's way of life.

There are formal rules and regulations written down in the corporation that outline how employees are supposed to do their jobs. There are also informal rules and ways of working that people develop within the corporation. Perhaps corporate management failed to adequately socialize the members of the sales department. Perhaps the sales department members failed to internalize the values, attitudes, beliefs, and norms of the corporation. As a sociologist, you see how valuable corporate training is.

The sales department may have also developed its own variations of the organizational culture and society. That is, it may have formed a **sub-culture**. Accordingly, its members may share a way of expressing camaraderie and corporate allegiance that the owners do not recognize. If you conducted an **ethnography** with the employees in this company, you may find **ethnocentrism**. You may find that the sales department thinks that it's much better than the engineering department. Of course, the engineering department thinks the same way about itself. Similarly, the owners think they know the best way for departments to work. In contrast, the departments think their way is the best way.

You might also find that departments do not share the same meanings for common words. The meaning of "being on time" varies across departments, managers, and owners. In other words, whether you're late for work or not is **culturally relative** to which department you belong. You might find that the meaning of "being productive" varies across departments and owners. It may be that the members of the sales department come in late in higher numbers because, in that department, employees work later hours or work at home or on weekends. You might also find different meanings of "sick." The sales department might define sick leave as personal leave, and may encourage employees to use their personal time if they have finished all their goals for that week or month. In doing so, they might save the company money by ensuring that time "on the clock" is productive time.

Understanding a group of people requires understanding the society and culture to which they belong. It also requires understanding people's sub-groups and sub-cultures. We can see from this example that there's not a perfect fit between culture and society. For example, the corporation wanted the employees to work hard and to be creative and productive, but the corporate culture also values employees liking their work and their coworkers, and respecting corporate leadership and authority. Sub-cultures and mainstream cultures often clash. Through this tension between society and culture, we get important social change. We will look at social change and tension between culture and society more thoroughly when we examine deviance later in the book.

Often social situations are so complex that just one of us would not be able to see or understand all of the possible social forces that influence it. Collective brainstorming will help us see the full picture. While there are many ways to brainstorm, here's one way we recommend.

Brainstorming

You need tools to drive your imagination. Sociology is pretty dry stuff if you don't use it, and you increase your personal and sociological skill by driving new ideas. Brainstorming is a way to do this. Gather a group of people who are interested in brainstorming the social factors that might influence some social situation.[3]

1. What is the question that you want to brainstorm? Record the initial and final drafts of the question.
2. All group members can give their ideas about the question. No criticism is allowed!
3. Write down everyone's ideas.
4. Keep providing new ideas until there are no more new ideas.
5. Rewrite the list of ideas, organizing them into categories. Delete duplications, but note how many times each idea was given.
6. Now repeat steps 2-4 to identify sociological concepts from this chapter that help make sense of the ideas listed in step 5.
7. How do these concepts help you understand this social situation?

TASKS

Understanding the sociological perspective is not an easy task. It is something that you rarely learn in just one course in sociology. Rather, it is something that you would learn across a lifetime. We think that the sociological perspective is a powerful tool that you can use throughout your life and career, regardless of your major. Here is an exercise to help develop your sociological perspective.

1. The federal emergency management department of a southeastern state hires you to work on an interdisciplinary team to increase the number of people who evacuate before a hurricane and to improve the evacuation process.
   - Offer a sociological explanation for why people do and do not evacuate.
   - Offer a sociological explanation for why large evacuations are difficult to manage.
   - What are some tensions between society and culture evident in your answers above?
   - What differences exist between sub-cultures regarding hurricane evacuation?

---

[3] Adapted from Brassard, M. and Ritter, D. (Eds.) 1994. *The Memory Jogger II*. GOAL/QPC: Methuen, Massachusetts. Pages 20-21.

Chapter 4. Groups and Organizations

QUICK START

In this chapter you will learn:

- How to distinguish between a group and an aggregate of people.
- To recognize the groups in which you are a member, and the ones in which you are not a member.
- The varying influences of different types of groups on social behavior.
- How to create a group from a collection of people.
- To understand the functions of different types of groups.

TERMS

Aggregate	A collection of people.
Group	Two or more people interacting who share similar interests.
In-groups	Primary or secondary groups with which we identify.
Out-groups	Primary or secondary groups with which we do not identify.
Primary Groups	Small, intimate, enduring groups such as your family. They provide a wide range of functions from protection, to emotional support, financial sustenance, and assistance.
Reference Groups	Any group from which we take our norms and values and to which we compare ourselves.
Secondary Groups	Larger, less intimate groups that form and exist in order to accomplish a goal(s); they often disband when the goal(s) is reached.

TOPICS

Plain and simple, sociology is about **groups**. Even people who know very little about sociology will indicate that sociologists study groups. That having been said, we need a practical view of groups. Let's take a minute to determine what groups are and what groups are not.

Envision yourself walking down a busy downtown sidewalk. Looking around you, there may be hundreds of people around you at any given moment. There are people walking and talking together, and some are simply walking beside one another. A variety of human conditions inhabit that sidewalk. Can we call the sidewalkers a group? After all, they are a bunch of people in one place at one time. Isn't that a group? Now a "bunch" may not seem very scientific, but most of us would probably imagine that a "bunch" of people in a common place would comprise a group. Why get any more complicated than that?

Well, we are not trying to get complicated, but a bunch of people sharing common space does not make a group. Instead, sociologists would call people who happen to be walking on the same sidewalk an **aggregate**. Now, talk about complicated, get this: "All groups are aggregates, but not all aggregates are groups." It takes more than a bunch of people being in the same place to form a group, and it's more than just the silly definition of a group that makes the difference.

So how are we to distinguish between a bunch of people that are and are not groups? What we need is a simple but powerful set of characteristics that help us distinguish groups from other kinds of bunches of people. Needless to say, sociologists have pondered the characteristics of groups over and over again. Sociologist Earl Babbie (1994)[4] offers a versatile set of criteria. He argues that an aggregate of people is a group if it has the following four main characteristics:

1. Shared Interests. Shared interests are more than having similar characteristics. For example, an aggregate of people would not be a group just because all of the individuals have brown eyes, or because they are all in the same place at the same time. Many people may have characteristics in common, but groups have shared interests. This means they have a "consciousness of kind." They have similar goals, values, attitudes, beliefs, behaviors, and experiences. For example, sharing the experience of discrimination and racism makes African Americans a group. This leads us to the second group characteristic.

2. Group Identity. Our colleagues in psychology make a great effort to ascertain the characteristics of individual identity. Identity is an important component in understanding individuals. However, identity is also important in understanding groups. Sociologists maintain that members of a group share a "we" feeling, a "consciousness of kind." Part of each individual's identity is an attachment to a group of people -- an awareness and feeling of belonging to something bigger than you. As an example, it's very likely that someone in your family has said to you, at one time or another, "That's not how we do it in our family." He or she may have said this when referring to some behavior that you engaged that went against your family's characteristics. In other words, you violated your family's shared sense of what and who the family is or stands for. Groups can have long-standing identities. They extend into the past, present and future. A group's identity answers the question, "What are we all about?" Group identity also connects the individuals in the group to a larger social order.

3. Group Structure. What is structure? This is probably the most difficult sociological concept to understand. So let's step back. For a building, structure is the girders that support the building. Let's try to apply that metaphor to a group. We need to look for the supporting elements of a group, such as norms. Remember that norms are rules for living. Norms are connected to shared values, beliefs, and attitudes. Norms tell us what behavior is appropriate and expected in any given situation. Other supporting elements of a group are roles and statuses. Remember that roles are sets of behavior expectations for a given social job or position (such as teacher, mother, friend, etc.). Statuses are the relative location or social prestige of each role. Norms, roles, and statuses provide structure to a group, giving us guidelines for interaction. Because of norms, roles and statuses, we don't have to reinvent the wheel for every social situation. We do not have to negotiate the rules of interaction in each situation. Hence, group structure supports collective action.

---

[4] Babbie, Earl. 1994. *The Sociological Spirit: Critical Essays in a Critical Science,* 2nd edition. Wadsworth: Belmont, California.

30

4. Interaction. Shared interests, identity, and structure will never come about if people do not act "among one another." Taking the word *interaction* apart tells us what we need to know here. The prefix "inter" means between and "action," of course, means to act. Interaction is the basic "stuff" of sociology and social life. Interaction is the glue that holds groups together. It simply not possible to have groups without interaction.

Now let's take a minute to consider these criteria once again. Is it possible to have a group without having all four of these? That's really a tough one! Perhaps it's better to say that a group is possible with varying amounts of each of these four criteria. We would be more likely to call an aggregate of people a group if the aggregate possessed all four of these criteria. But groups do exist and persist at varying levels of each of these criteria.

Next we need to consider why groups are important. Groups are important because they serve several functions for the societies in which they exist. To understand group functions we must first explain some types of groups. We have **primary** and **secondary groups**, and these can serve as **in-groups** and **out-groups**.

Primary groups are small, intimate, enduring groups such as your family. They provide a wide range of functions from protection, to emotional support, to financial sustenance and assistance. Secondary groups are larger and less intimate. They form and exist in order to accomplish a goal(s), and often disband when the goal(s) is reached. Examples include your coworkers, members of a civic organization, and students in a class. Usually members of secondary groups do not provide emotional support or financial assistance, though they may indirectly assist members financially by serving as professional networks.

In-groups are primary or secondary groups with which you identify, or aspire to be like. People imagine how an in-group would react to their behavior in any given situation. Violating in-group norms makes people feel shame and guilt. People engage in behaviors that produce positive in-group reactions and avoid those that produce negative in-group reactions. For example, the group of "cool" high school students with whom a teenager wants to associate represents an in-group to him. Similarly, a gang represents an in-group for its members. In contrast, a rival gang represents an out-group, as would a group of "nerdy" people that a teenager shuns. People engage in behaviors that produce negative out-group reactions and avoid those that produce positive reactions. In short, people evaluate themselves against in- and out- groups. The groups of people we identify with and those we distance ourselves from shape our sense of self.

Reference group is a general term that sociologists use to denote in and out-groups. For example, say you are considering marrying someone of a different race than yourself. How would your in-groups respond? How would your out-groups respond? Is the pressure to belong enough to keep you from marrying this person? Let's say you decide to marry. Now you are thinking of hyphenating your name. What problems do you anticipate from your primary and secondary groups? Now, you and your spouse are considering moving away from your home town in order to take new jobs. How would your reference groups respond? What would you do in order to maintain your relationships with your reference groups if you did move?

31

TOOLS

So how can we link our knowledge of groups to an applied sociological tool? This is pretty straightforward. If we know what conditions make a group, then we know how to build a group, which is a valuable skill in today's world. As an example, suppose a manager of a new department hires you to help a new team of employees work together. Your task is to turn an aggregate of people into a group. Let's create a group.

Tool	What Is It? How Do I Do It?
Create a group[5]	Identify an aggregate of people who are not already a group. This collection of people may, but need not, occupy the same general location. E-groups, for example, are possible, too.
Establish interaction	Naturally the survival of the newly formed group would strengthen if its members continued interacting. What are the goals of the group? Do the group members need to continue interacting to meet these goals? Can we create goals that require the group members to continue interacting? • Determine the group's purpose • Determine the core values for the group. • Establish ways to interact on a routine basis. • Create means for communication.
Produce shared interests	Create or determine shared interests. Our challenge would be to purposefully create situations in which people recognize their existing shared interests and situations in which they develop new shared interests. • Identify what interests the group already has. • Identify what interests we want the group to have.
Establish a collective identity	We will also need to answer the question, "Who and what is this group?" Can we create situations in which the individuals in the group acknowledge collective ownership for the group's activity? To reinforce group identity, we will need to lay down the boundaries of where this group starts and stops, and where other groups start and stop. Reference group behavior can be very helpful here. • Decide who "we" are. • Determine symbols and rituals that signify your group's boundaries.
Determine group structure	Naturally this group will need girders to support action. What are the group's functions and activities? Who will do this work? Will there be a division of labor? Does the group require a leader? Can we construct sets of norms and roles that flow from these functions and activities? Are some roles deemed more important than others? • Establish the norms, the rules for the group. • Decide how you're going to make decisions. • Determine the roles and statuses -- how and who will lead? • Establish how you bring new people into the group and how people leave the group.

[5] Portions of this section were adapted from Brassard, M. and Ritter, D. (Eds.) 1994. *The Memory Jogger II*. GOAL/QPC: Methuen, Massachusetts. Pages 150-151.

TASKS

1. You are a clinical sociologist working with an extended family that is divided over whether the grandparents should move from their home to a nursing home. How can you use what you know about groups to bring this family back together again?

2. One of your college professors assigns you and five other classmates to work together on a class project. Is this a group? If so, how so? How can the sociology of groups help you accomplish your class project?

QUICK START

In this chapter, you will learn:

- How to recognize deviance.
- How to apply the functions of deviance to a social situation.
- Why deviance is not always negative.
- How groups control social action.

TERMS

Control Theory	The sociological explanation of deviance that maintains people conform due to effective social control and self-control.
Deviance	Behavior that departs from the norms of a group.
Differential Association Theory	The sociological explanation of how people learn to act normative or deviant from people around them.
Formal Social Control	Social regulation of behavior through laws.
Functions of Deviance	The purpose that deviance fills in society, including producing change and reinforcing normative behavior.
Informal Social Control	Social regulation of behavior through social pressure and customs.
Labeling Theory	The sociological explanation of deviance that maintains deviance occurs when people define characteristics and behaviors as deviant and then respond to them accordingly.
Master Status	A role that overshadows all of a person's other roles and statuses.
Negative Social Sanctions	A punishing social response to deviant behavior.
Normative Behavior	Behavior that follows norms of a group at a given time and place.
Pluralism	A culture with multiple subcultures and ways of seeing the world.
Positive Social Sanctions	A rewarding social response to normative behavior.
Primary Deviance	Daily life departures from the norm in which many people engage.
Secondary Deviance	Extreme departures from the norm that convey a commitment to a lifestyle of norm breaking.
Self-Control	Individual regulation of behavior, due to internalizing a society or group's expectations regarding values, attitudes, beliefs, and behavior.
Social Control	Social regulation of behavior.
Socially Constructed	Patterns of behaviors, expectations, attitudes, and beliefs that people create over time through interaction with other people.
Stigma	A mark of social disgrace, changing a person's self-concept.
Stigmatize	The process by which people apply a stigma to a person or group.
Structural Strain Theory	The sociological explanation of deviance that maintains people act deviantly depending on whether they accept cultural goals and perceive access to the structural means to reach those goals.

TOPICS

Hardly a day goes past when we don't hear someone say, "Wow, that's weird!..." Usually we say this in response to someone engaging in behavior that we see as wrong or different, outside the bounds of acceptable activity. In other words, they were acting deviantly. But what we define as weird or deviant changes across people, place, and time. For example, 20 years ago it would have been weird to see a child wearing a helmet while he or she rode a bike. Today, we think it is weird if children do not wear helmets. Similarly, we consider it normal to see babies in diapers, but would consider it very weird to see a 40-year-old adult wearing diapers. And it is considered deviant to yell loudly while at a movie theater, but it is expected while in a sports stadium.

So, you see, what is deviant depends on the situation you are in, the people who are there, and the historical time. Groups of people decide, through social interaction, what is deviant. These definitions of **deviance** change. Sociologists define deviance as the departure from **normative behavior**. Let's spend some time on this definition of deviance.

To get to the root of the sociological definition of deviance, we need to reflect on cultural relativity for a minute. You should remember that things have meaning only within the context of the culture in which they exist. When we say something is deviant, we mean that it is deviant relative to other behaviors in that setting or culture. By now you know that all societies and groups have norms or rules for behavior. These rules are tied to the group's values, attitudes, and beliefs. These rules are reinforced over time to provide some baselines for human action. It is not the action itself that is deviant, but, rather, how the action is seen relative to the norms for that situation, actor, and audience.

For example, imagine a family vacationing on an American beach. They see a group of young women walking along wearing thong bikinis. The parents, thinking this was a "family" beach, find the young women's behavior deviant. However, other beach goers may not find the wearing of thong bikinis deviant at all. In fact, vacationers from other countries may find it unusual that women wear tops on American beaches. So you see, behavior itself, in this case the wearing of a thong bikini, is not deviant. Rather, deviance lies in our perceptions. A behavior only becomes deviant if the actor or audience in that situation defines it as such. Deviance is, as sociologists say, **socially constructed**.

We also want to illustrate that context shapes whether we consider a behavior, value, belief, or attitude deviant. Take underage drinking, for example. In the United States, we consider a 16-year-old ordering a beer in a bar deviant, as do we any bartender who serves a 16-year-old beer. In England people consider these same activities normal. Norms change across time and place. Behaviors, values, beliefs, and attitudes can only be understood within the situation in which they occur. We could take this exercise a step further and ask the question, "Which situation is right and which is wrong?" Now of course, right and wrong suggest values. When we call something strange or weird, we often not only attach a description to it, but also a value. Whose values are they? What a society defines as right or wrong, normal or deviant, reflects the interests of those in political, economic, and social power. That is, norms and laws reflect the wishes of those in power. For example, it is legal for homeowners to receive tax deductions for owning property but no such provision exists for people who rent housing. Similarly, we enforce laws in ways that advantage those in power. For example, we prosecute more people arrested for possessing crack cocaine than people arrested for possessing powder cocaine. Money or power buys the privileged excellent legal representation should arrest or prosecution occur, which often enables them to resist deviant definition or treatment.

Both of the above examples indicate that we collectively define, or socially construct, normative and deviant realities. An understanding of the relative nature of deviance leads us to an awareness of an ever-increasing pluralistic environment. **Pluralism** comes from the word *plural*, meaning "many." "Many of what?" you might ask. Many different collective realities, we add. So if we live in a multicultural environment, we should expect multiple collective definitions of normative and deviant.

Don't confuse deviance with unhealthy behavior. Sure, some individuals and groups may find all deviant behavior unhealthy. But with a sociological perspective, we recognize that deviance also provides some extraordinarily positive outcomes too. As societal needs change, things that we defined as deviant may become the norm. The tension between culture and society sometimes gives way to a change in cultural norms. When this happens deviance takes on the important function of providing a vehicle for needed social change. Another function of deviance is to demonstrate, by its sanctioning, the baselines or norms for living. These functions seem quite contradictory to one another. But they operate in a precarious balance with one another. For example, on one hand, deviance provides the basis for structured social activity, while on the other hand, deviance heads off stagnation by providing a means for social change. In our rapidly changing technological environment, deviance represents a critical part of our social life.

Some of the other **functions of deviance** include:
1. It affirms cultural values.
2. It generates and sustains morality.
3. It clarifies moral boundaries.
4. It encourages discussion of issues.
5. It promotes social unity.

TOOLS

Labeling Theory

Labeling theory expands on the idea that characteristics and behaviors are not inherently deviant. Instead, people must define characteristics and behaviors as deviant and then respond to them accordingly. In turn, people respond to the label of "deviant." **Primary deviance** refers to common, minor norm violation like picking your nose or farting in public, and the petty theft that adolescents often engage in. Primary deviance is temporary, and not central to one's identity or to others' impression of people. In contrast, **secondary deviance** refers to severe deviance, such as murder, or sustained deviance, such as someone who becomes a career petty thief. Secondary deviance results in a label as "deviant" that sticks, becoming a **master status** or **stigma** central to one's identity and others' impression of people. In effect, people become **stigmatized** by secondary deviance. For example, when we label a person as deviant for murdering his or her children, we only see the murderer as deviant, disregarding the other, non-deviant roles that he or she may play such as sibling, friend, or co-worker. Similarly, early Americans labeled women who engaged in adultery or supposed witchcraft deviant for the rest of their lives, often resulting in social isolation or death.

Structural Strain Theory

Like labeling theory, structural strain theory, a variant of functionalism, offers an explanation for why people act deviantly. Functionalism maintains that social structures more or less determine people's behavior. Structural strain theory applies this approach to deviance. Robert Merton (1938) explains that the access to and acceptance of cultural goals, and the structural means to reach those goals, determines whether people act deviantly or not.[6] Some people conform. These people accept the cultural goals, such as being wealthy, and accept the structural means, such as earning a college degree. In contrast, some people reject either or both a society's goals and means to achieve them. We consider all of these people deviant.

Some people accept a society's goals but use alternate, innovative, means to reach them. Often, people who innovate new means do so because they are blocked from using accepted means such as education, training, and jobs. To reach the accepted goals of wealth and happiness, innovative people use means such as robbery, burglary, selling drugs, or selling sex. In contrast, other people accept the means but disregard the goals. These are ritualists who, for example, go to college because it is expected of them, not because they want to be successful in a particular career. Still other people retreat from both society's goals and means to achieve them. Many Americans see homeless people as the classic example of a retreatist (not realizing that many homeless people are structurally blocked from American goals and means to attain them). Finally, some people rebel against a society's social structure, rejecting both society's goals and means, and replacing them with their own. Rebels include members of various social and cultural movements and revolutions, such as the women's movement and the civil rights movement of the past, and the current environmental and gay rights movements.

Control Theory

Control theory, picking up where structural strain theory leaves off, explains how people come to reject a society's goals or means to attain them. People conform due to effective **social control** and **self-control**. **Formal social controls** and **informal social controls** serve to either encourage people to follow the social norms of a given time or place (**positive social sanctions**) or to deter people from violating the social norms of a given time or place (**negative social sanctions**). For example, we encourage adults not to break laws by providing a host of positive

---

[6] Merton, R.K. 1938. Social Structure and Anomie. *American Sociological Review* 3(6):672-682.

sanctions, including social, political, economic, and religious freedom. We deter people from breaking laws with the threat of various negative sanctions such as fines, denial of college funding, removal of voting privileges, higher insurance costs, and prison time.

Control theory argues that people who act deviantly do so because they lack either social or self-control. To understand how this would happen, let's first look at how we internalize social controls and become self-controlled. Through socialization, we attach ourselves to reference groups, and with time, learn to anticipate their reactions. For example, over time, people become attached to various reference groups such as their families, friends, churches, schools, work, and community. Being involved in these groups motivates us to make the group's norms our own. In doing so, we develop identities, beliefs, values, and attitudes that are closely linked to our group attachments. In turn, these groups give people opportunities to become involved in socially legitimate activities. People who violate norms either lack attachment to legitimate groups or they do not sufficiently anticipate the reactions of these groups. To prevent this from happening, social control theory would advise parents to get children involved in legitimate group activities, and for adults to be active members of their communities.

Differential Association Theory

Differential association theory goes a step further than social control theory and offers an explanation of how people learn to act deviant from their attachments to different (or illegitimate) groups. Differential association theory states that people learn to be deviant from the people around them. If the people around you participate in deviant acts, you learn to value these acts and learn how to do them. If these people are a reference group for you, they can impose positive and negative sanctions to get you to conform to their expected behavior. The more often you interact with these deviant people, the more likely you are to participate in deviance too. Once you begin engaging in deviance, you may learn how to perform more severe deviant acts, either through experience or through observing others. For example, you may start out stealing candy and progress to stealing clothing, cars, or money. To use the language of labeling theory, a primary deviant may also learn from secondary deviants how to prevent being caught and labeled. For instance, it is commonly known that many imprisoned people learn how to commit more serious crimes from associating with other prisoners.

TASKS

1. A small group of employees at a car factory are considering unionizing. Use what you know about deviance to identify the problems that they can anticipate from the owners, managers, and fellow employees.

2. A local church hires you to design a program to discourage drug use among its youth. Use the topics and tools from this and the previous chapter to identify components of your proposed program.

3. Create a set of rules of conduct for patrons of a new community library.

Chapter 6. Diversity and Inequality

QUICK START

In this chapter, you will learn:

- An awareness of diversity.
- The sociological meaning of gender, sexuality, age, race, ethnicity, and social class.
- How the social forces of gender, sexuality, age, race, ethnicity, and social class shape us and society.
- How people create and reproduce social divisions and inequality.

TERMS

Age	Socially defined expectations, obligations, and statuses ascribed to persons at chronologically different times in the life cycle.
Diversity	The ways in which persons are perceived as different -- for example, race, ethnicity, class, and gender.
Ethnicity	A group of people who share the same cultural traditions.
Gender	Roles that persons of different sexes play.
Glass Ceiling	The lack or denial of advancement opportunities for minorities and women in some professional careers.
Inequality	The socially constructed outcome of unequal resources (power, prestige, and wealth) among groups of people.
Life-Chances	The probability that certain life events will occur, often based on one's position in the social structure.
Race	A group of people who share a common historical and cultural realities based on skin color.
Self-Fulfilling Prophecy	The interaction between one's self and others in which you become what you expect others want you to be.
Sex	The biologically determined differences between women and men.
Sexuality	The physical attraction that people feel for one another.
Social Class	A group of people who share a common level of wealth.
Stereotype	An unrelenting image of a group, whether true or not.

TOPICS

The concern over diversity among people is an extremely important part of applying sociology. At the societal and multicultural levels, the issue of **diversity** is taking on growing importance. As global migration increases, it becomes more likely that you will interact with people quite different from you. The Internet and rapid modern transportation have had a similar impact on global interactions. Pluralism has become a way of life for the 21st century.

Let's reflect on how people become different in the first place. We know that we socialize each other. We acknowledge the importance of individuality in this process, but as sociologists we look at how social forces, external to individuals, influence personality and behavior. Taking in the "stuff" outside of us helps shape who we are. In this sense, everyone is unquestionably different and yet similar, in terms of the contexts in which they experience socialization across their lifetimes. Following this sociological perspective, if we identify and understand the social forces involved, then we can better understand what patterned similarities and differences exist among people, and why they exist. This understanding, theoretically, should improve our interaction with each other.

So where do we start? Good question. Many social forces shape who we are. Below we look at some of the most important social forces that make people similar and different in our society. To learn more about these and other social forces, take more sociology courses!

Gender

Women are different than men. Wow, that's another one of those obvious sociological statements! As simple as it sounds, it's extremely powerful. Now we know that men and women are biologically different. Sociologists refer to the characteristics that make up biological differences as **sex**, as in the "battle between the sexes." In contrast, the term **gender** refers to the different social roles that men and women play. We learn these different social roles from the society and culture in which we live. For example, who usually provides the majority of child care? Who usually provides the majority of household income? Who usually changes his or her name upon marriage? How do we name children? What kinds of toys do boys play with? What kinds do girls play with? We could go on and on here. Suffice it to say that most of the differences between men and women are due to gender, not sex.

The roles and statuses that accompany gender have a powerful impact on who we become. While it would be wrong to categorize all men as the same and all women as the same, it is likely that men will have more similarities in social roles with other men, and women will have more similarities in social roles with other women, particularly if they have been rasied in the same society and culture.

Below, we display the first row of a table, which we call a "diversity table," to which we will add each social force, beginning with gender. As we add social forces, the diversity table will illustrate the complexity of difference.

Social Forces	Examples	
Gender	Woman	Man

Sexuality

Many people confuse gender, sex, and sexuality. They think gender and sex are the same thing, and they think sexuality is the same thing as sex and gender. Above you learned what distinguishes sex from gender. Now, let's look at sexuality. Sexuality refers to the physical attraction that people feel for others. Some people feel physically attracted to people of the same sex (homosexuality), or to the other sex (heterosexuality), or to people of both sexes (bisexuality). It is not that simple, though, because people use the term *sexuality* to refer to a person's sexual identity, desire, and behavior. People's identities, desires and behaviors may not match. That is, people may think of themselves as homosexual (identity) and feel sexually attracted to people of the same sex (desire), but never engage in sexual relations with people of the same sex (behavior). Similarly, people may think of themselves as heterosexual, but feel sexually attracted to people of the same sex, and engage in sexual relations with people of the same sex. Bottom line: Sexuality is complicated.

Now, we know that people of different sexualities act differently. Many people make this distinction bigger than it really is. Many people think that sexual differences carry over into every aspect of a person's life. They think that, by definition, homosexuals, heterosexuals, and bisexuals hold different attitudes, beliefs, values, and behaviors about family, work, careers, religion, health, and so on. Actually, people of different sexualities are much more similar than they are different. For example, most people, regardless of sexuality, long to be loved and to love, to have a family, to live in a nice home and neighborhood, and to be successful at work. Most people, regardless of sexuality, have friends of both sexes, like to go on vacation, worship a God, have problems with their cars, and so on. However, because people make such a large distinction between people of different sexualities, let's add them to our diversity table.

Social Forces	Examples			
Gender	Woman		Man	
Sexuality	Homosexual	Heterosexual	Homosexual	Heterosexual

Age

Are you in the net generation? A generation X'er? A baby boomer? While sociologists debate the labels, they tend to agree that the cohort in which you were born shapes how you see the world. People in an age cohort tend to have similar life experiences. They take in what is "going on" historically at each point in their lives. Hence, we see cohort effects in people's attitudes, beliefs, values, and behavior. For example, older Americans today more strongly oppose same-sex and inter-racial marriages than younger Americans. Americans who reached adulthood during the Vietnam War embrace more liberal and socially conscious beliefs and values. Adults of the MTV generation, as a group, seem the most concerned about environmental issues. This is not to say that people do not change and learn new ways of being and doing, but rather that people in an age cohort tend to hold similar norms and values, a similar age-related way of life. So let's add age to our diversity table.

Social Forces	Examples							
Gender	Woman				Man			
Sexuality	Homosexual		Heterosexual		Homosexual		Heterosexual	
Age	25-40	41-56	25-40	41-56	25-40	41-56	25-40	41-56

Notice what is happening in our table. When we started, we attributed the differences in people to gender: women are different than men. Now, see this at a higher level of complexity. Homosexual women age 25-40 experience a different social reality than homosexual women age 41-56, and both groups differ from what heterosexual men age 25-40 experience.

Ethnicity and Race

People from different cultures act differently. Again, this is simple but powerful. By now, you probably feel comfortable with this. You've seen the cross cultural outcomes of different ways of life in the media. You may have experienced different cultures in other societies or different subcultures within our society. Members of ethnic groups, such as the Irish and Jewish, share unique cultural traditions and histories. Sociologists argue that the same is true of race.

For example, generations of African Americans share a similar experience of discrimination and racism in the United States. They share similar ancestors and many marry and raise families with other African Americans. As a group, African Americans maintain a shared identity and cultural heritage through dress, music, food, religion, dance, art, literature, language, and community. The same can be said of other racial groups, such as Asian Americans, Native Americans, Latino Americans, and European Americans.

People of the same race share historic and cultural traditions and experiences, as do members of ethnic groups. Not though because of any inherent differences in the biology of skin color, but rather because of the meaning we socially construct for skin color or race. In other words, racial differences exist because we make them exist.

Now our table increases in complexity, because race and ethnic groups within the same culture may have different experiences from other groups.

Social Forces	Examples																															
Ethnicity	Jewish																Irish															
Race	Asian-Am.								African-Am.								Asian-Am.								African-Am.							
Gender	W				M				W				M				W				M				W				M			
Sexuality	H1		H2		H1		H2		H1		H2		H1		H2		H1		H2		H1		H2		H1		H2		H1		H2	
Age	25-40	41-56	25-40	41-56	25-40	41-56	25-40	41-56	25-40	41-56	25-40	41-56	25-40	41-56	25-40	41-56	25-40	41-56	25-40	41-56	25-40	41-56	25-40	41-56	25-40	41-56	25-40	41-56	25-40	41-56	25-40	41-56

H1=Homosexuality   H2=Heterosexuality

Social Class

Social class has a huge impact on social behavior. Most people self-identify themselves as members of the upper, middle, or lower class, and this identification influences how they see the world, their values, beliefs, and attitudes. For example, research shows that parents in the lower class teach their children to mind their teachers and to obey authority. In contrast, parents in the middle and upper social class teach their children to question authority and to think critically about what people tell them. Hence, the terms upper, middle, and lower class (and gradations within) correspond to different lifestyles.

Through group interaction, most societies separate people into different social classes. For example, people with economic wealth can send their children to the top schools, patronize the best restaurants, purchase memberships in exclusive social and recreational clubs, live in gated communities, own businesses and stocks, hold the best jobs, and exert influence in political campaigns. People without economic wealth are blocked from similarly participating in these same opportunities, and instead live very different lives. As a result, the social circles of the upper class and the lower class rarely overlap. For the most part, people with wealth interact with other wealthy people, and people without wealth interact with people without wealth. This separation helps to maintain the boundaries of each social class, and each class's privileges, or lack thereof. Below, we add social class to our diversity table. In doing so, the table becomes so complex that we need to show separate tables for each social class.

Upper Class

Social Forces	Examples															
Ethnicity	Jewish								Irish							
Race	Asian-Am.				African-Am.				Asian-Am.				African-Am.			
Gender	W		M		W		M		W		M		W		M	
Sexuality	H1	H2	H1	H2	H1	H2	H1	H2	H1	H2	H1	H2	H1	H2	H1	H2
Age	25-40 / 41-56	25-40 / 41-56	25-40 / 41-56	25-40 / 41-56	25-40 / 41-56	25-40 / 41-56	25-40 / 41-56	25-40 / 41-56	25-40 / 41-56	25-40 / 41-56	25-40 / 41-56	25-40 / 41-56	25-40 / 41-56	25-40 / 41-56	25-40 / 41-56	25-40 / 41-56

Middle Class

Social Forces	Examples															
Ethnicity	Jewish								Irish							
Race	Asian-Am.				African-Am.				Asian-Am.				African-Am.			
Gender	W		M		W		M		W		M		W		M	
Sexuality	H1	H2	H1	H2	H1	H2	H1	H2	H1	H2	H1	H2	H1	H2	H1	H2
Age	25-40 / 41-56	25-40 / 41-56	25-40 / 41-56	25-40 / 41-56	25-40 / 41-56	25-40 / 41-56	25-40 / 41-56	25-40 / 41-56	25-40 / 41-56	25-40 / 41-56	25-40 / 41-56	25-40 / 41-56	25-40 / 41-56	25-40 / 41-56	25-40 / 41-56	25-40 / 41-56

Lower Class

Social Forces	Examples															
Ethnicity	Jewish								Irish							
Race	Asian-Am.				African-Am.				Asian-Am.				African-Am.			
Gender	W		M		W		M		W		M		W		M	
Sexuality	H1	H2	H1	H2	H1	H2	H1	H2	H1	H2	H1	H2	H1	H2	H1	H2
Age	25-40 / 41-56	25-40 / 41-56	25-40 / 41-56	25-40 / 41-56	25-40 / 41-56	25-40 / 41-56	25-40 / 41-56	25-40 / 41-56	25-40 / 41-56	25-40 / 41-56	25-40 / 41-56	25-40 / 41-56	25-40 / 41-56	25-40 / 41-56	25-40 / 41-56	25-40 / 41-56

As you can see in the tables above, understanding diversity can be complicated. People are not just women or men, not just black or white. We are Jewish, African American, heterosexual men, age 25-40 of the upper class. Or Irish, Asian, homosexual women, age 41-56, of the middle class. Each dimension of diversity adds another lens and layer of experience. Understanding these dimensions will help improve relations between people in any setting. Understanding these dimensions can help you explain why and how social problems occur.

TOOLS

Why is diversity so important to sociologists? How can we use knowledge about diversity to explain social problems? The concepts discussed in this chapter can provide you with a starting place for understanding how **inequality** causes many social problems. In our society, and in most other societies across history, we distribute, or stratify, social resources according to people's gender, sexuality, age, race, ethnicity, and social class. With any social problem or question, you may want to look at who has social resources (power, prestige, and wealth), who doesn't, and how we maintain these divisions. Let's walk through an example.

Suppose you are a human resources manager for a national department store. An employee in one of the stores files a complaint that she was overlooked for a promotion to a management position. During your investigation, you review the files of all the employees eligible for a promotion in the last ten years and all those considered for a promotion. You record the gender, age, race, and ethnicity of each employee eligible and each employee considered for promotions. You plan to use this data to see if the percentage of people eligible of each gender, age, race, and ethnicity group matches the percentage of people promoted of each gender, age, race, and ethnicity group. You would like to also analyze the sexuality and social class of each employee, but those data are not available.

The data indicate that the department store promotes a disproportionately high number of white men age 30-50. This leads you to look at the characteristics in each store of the people who make promotion decisions. You find that the characteristics of the decision makers match those of the promoted people. These findings suggest that a **glass ceiling** for women, racial and ethnic minorities, and older employees exists in your company due to discrimination and prejudice. One of the results of this discrimination and prejudice is that the minority and women employees have less promising **life chances**, which then influences their immediate and extended family's life chances, as well as the life chances of people in their communities who identify with them. Consequently, you see the discrimination and prejudice in your organization as a social problem, not just an organizational problem.

You take these findings to the vice president, CEO and the board of directors. You make the following recommendations. First, you recommend a more thorough review of the data, including investigating the potential for discrimination and prejudice at all levels of employment. You suspect that the department store grooms a disproportionately low percentage of minorities and women for management. Subsequently, a lower percentage of minorities and women become eligible for promotion. Second, to address the problems of prejudice and discrimination in the company, you recommend the following:

1. The development and implementation of a professional mentoring program that targets minorities and women. This recommendation includes measuring the outcomes of this program to determine the program's success. You suspect that a **self-fulfilling prophecy** may be operating. The minority and women employees think they lack qualifications for management positions. The company administrators hold similar beliefs, perhaps because of prejudice. As a result, the women and minority employees do not seek additional training, nor do the company administrators encourage them to so. Then when management positions open, the women and minority employees do not apply for these positions, nor do the company administrators invite them to do so. Because minority and women employees do not plan for or seek management positions, the administration's supposition is supported; thereby legitimizing, in their minds, their actions and beliefs, which they subsequently do not define as prejudiced or discriminatory.

2. The development and implementation of a diversity awareness program, or a contract with an outside vendor specializing in such programs. The diversity awareness program would target management and administrative level employees, including the vice president, CEO, and board of directors. You recommend that the company track this program to determine its success.

Speaking of diversity awareness programs, let's take a look at how you might conduct one.

Tool	What is it? How do I do it?
Mini-ethnography	Define a target area for investigation. For this tool we recommend that it be a small social environment -- your neighborhood or your company, for example.
Getting started	Start with and respect the assumption that differing ways of life exist (you don't need to like them or agree with them). For the moment, erase or at least hold in check, to the best of your ability, any preconceived attitudes that you might hold about people different from you.
Observation	Systematically "walk around" the environment; look and listen for: a. Patterns of people who are interacting. b. The characteristics of these people. c. The location of different groups of people doing, saying, or acting in patterned ways. d. Look and listen for patterns of words and symbols. Are there any apparent patterns emerging that reflect who and what these people are doing or saying? When you get a chance, write down what you've heard and seen.
Identify Key Informants	4. In most social environments there are "key informants." These are people who "know what is going on" because of their status, longevity, seniority, etc. a. Locate key informants by asking people in the social environment who they would recommend as someone who understands what's going on here. b. Ask several groups this question. Make mental notes of those persons who are recommended. Patterns should develop. When you get a chance, write down what you've heard and seen.

	c. Make appointments to talk to some or all of these key informants. Your goal is to find out the level of diversity in this group. You want to inquire as to the patterned similarities and differences among the people in this area. Write down what you've heard and seen.
Summarize	Write a brief summary of the diversity that you learned of in your observations and discussions with key informants. List the different ways of life that you would attribute to diversity.
Validation	Take your summary and your list to the key informants. Do they agree with your observations and conclusions?

In concluding this chapter, we'd like to recommend that you apply the concept of diversity with caution. First, avoid **stereotyping** people with invalid labels. This simply means that what may be generally true for a group may not be true for an individual in that group. Second, remember (and this may surprise you coming from sociologists) that at the micro level, individuals really do have personalities! It is often difficult to sort out cultural norms from individual personality traits. Finally, one way to cautiously approach diverse groups is to "know what you don't know." By this we mean that you need to be aware of the holes in your understanding of people. Knowing where these holes lie may keep you from falling down one.

TASKS

1. Your employer wants to encourage understanding and respect between race and ethnic groups. Propose three different programs that your employer can put in place at work to reduce racism and discrimination.

2. Three local ministers of different churches want to bring their congregations together into one united non-denominational parish. The members of each church differ in race, ethnicity, and social class. What problems should the ministers be ready to confront?

3. You work as a data analyst for the police department of a large Northeast city. Five years ago, the police department began documenting and tracking hate crimes. The hate crime data show that most are committed by men aged 18-25 against gay men, and that the number of hate crimes against gay men is increasing dramatically. Using the sociology of gender and sexuality, determine why this is happening so you can help develop prevention and educational programs.

Chapter 7. Economy and Work

QUICK START

In this chapter, you will learn:

- An overview of work and occupations that may help you with your future.
- How work has an impact on your personal and social identity.
- How life-long learning and work are integrated.
- A snapshot of the changing view of work.

TERMS

American Work Ethic	The belief that "hard work" is central to the American way of life and characterizes American society.
Automation	Technological control of production that minimizes the need for labor.
Blue-Collar	Jobs, usually requiring trade skills, in which people often wear uniforms, such as plumber, mechanic, wait staff, electrician, police officer, cashier, etc.
Capitalism	A form of economic organization in which the means of production are owned privately. There is no redistribution of wealth and free markets and profit incentives predominate.
Class Inequality	The socially constructed outcome of unequal resources (power, prestige, and wealth) among groups of people holding different economic positions.
Comparable Worth	The idea that positions that differ in title or job description but that require similar skill and responsibility should pay similarly.
Credentialism	The social phenomenon of employers demanding higher educational and training levels among job candidates for jobs that do not require such skills levels to perform; and, in turn, the race among job candidates to acquire additional education and training, the completion of which provides more degrees and certificates that the job candidates can add to their resumes and, via acronyms, behind their names.
Cross-Train	A strategy for training, structured to provide a person the ability to perform multiple skills across a set of different job roles.
Deskilling	Reducing the level of skill required to perform a job, reducing to smaller aspects of one job) to increase control of workers by management and reduce labor costs.
Economy	The social institution regulating how goods and services are traded.
Life-long Learning	Continuing to learn new skills and tools across your lifetime.
Occupations	Groups of jobs with similar responsibilities and skill sets. Such as the occupation of teaching, which includes jobs such as elementary school teacher, gym teacher, college instructor, and full professor.
Occupational Segregation	When an occupation contains mostly men or women, such as nursing, elementary school teacher, secretary, and day care workers (mostly women) and steelworker, farmer, economist, politician (mostly men).
Paternity Leave	Excused work absence for men after their child's birth or adoption.

Profit-Sharing	A distribution of profits structured such that individual designated workers within the system get a designated portion of the profit of the company as a whole.
White-Collar	Professional jobs, usually requiring college degrees, and in which people often wear suits or dress shirts, such as manager, doctor, lawyer, accountant, teacher, etc.
Work	The process through which we transform our own personal energy, through real or imagined interaction with others, to do something productive for society.

TOPICS

Your may ask yourself, "Why am I in school?" Your parents may ask the same question. If you are a returning student, your kids may ask this question, too! You may frequently answer, "to get a good job or to receive a promotion or raise," or "to change jobs." Those responses make sense. The information and skills you learn in college ought to improve your occupational success and participation in society. We think that sociology offers important tools in helping you achieve these goals. In fact, we think these goals warrant taking some time to look at the nature of **work** and **occupations**. Here's what we mean.

Sociologists think of work and occupations as another social institution, more generally known as our **economy**. Why do sociologists consider work and occupations an institution? Because people need to "produce" in order to survive. All of us rely on others to produce, that is to do their jobs. We assume there will be food at the grocery when we go, that our phone will work, that our money in the bank is safe, that our kids are being cared for in day care or school. Since we are so interdependent, we need to establish norms and sanctions regulating the production and distribution of resources. Hence, we have a social institution of economics.

First, what is work? From a sociological perspective, work is the process through which we transform our own personal energy, through real or imagined interaction with others, to do something productive for society. How do work and sociology go together? We'll remember that sociology is all about human interaction. Can you think of any work experience that doesn't include some level of human interaction? Sure you could work alone, which may become more prevalent as more people telecommute. Yet, when it's all said and done, the work that you do alone impacts others. In some way or another, people either buy, sell, or use the "product" that you create alone. Going a step further sociologically, the work that you perform is not only in interaction between yourself and other person(s), but also an interaction between your unique creative abilities and social set of expectations and obligations. What does that sound like? To a sociologist, it sounds like roles.

Your expectations of yourself and your perception of other people's expectations of you strongly influence the work that you do, your experience of work, and the importance of work in your life. For example, all of us share an idea of what it means to be a teacher, a lawyer, a mother, and so on. We learn, through socialization, different work roles and what to expect of people performing those roles. Further, patterns exist across societies, age, gender, race, and religion in work, the experience of work, and the importance of work. For example, we expect and prepare men to work and to make work a central aspect of their identity. In contrast, we expect and prepare women to choose whether to work. Women can choose to "not work" and instead to care for homes and children (which is work, it just usually isn't paid). Moreover, women can go in and out of the work force across their lives, at times choosing "not to work" in order to provide child or family care.

Imagine a man choosing not to work and instead caring for his wife, children, and home. Imagine men going in and out of the work force across their lives to care for children and family. How would people respond? One word: negatively. His friends and extended family would likely devalue and ostracize him, thinking him less than a man. His wife may not even fully support this occupational choice, having expected to marry a man who can and does provide for his family. Subsequently, most men do not even consider "not working" and most women would not want them to either. In fact, men so strongly internalize the expectation to work that when their employers offer **paternity leave** most men refuse to take it.

In the U.S. today, men, and a growing number of women, derive great meaning in their lives from their work. We define ourselves by our work. We work to enhance our personal identity and social status. Across history or societies, work has not always been so meaningful as today in the U.S. In ancient Athens, people considered working for pay undesirable. Only peasants worked. Instead, people spent time thinking. In the past, Hebrews considered people sentenced to work as a result of Adam and Eve's mistakes. Early Christians performed "good works" to obtain God's blessing. Early Protestants expanded this idea to serve God by performing any work, making it a moral duty to work. People who reached economic success were thought to be in God's good graces. These Protestant beliefs drove what became the **American work ethic**, largely shaped by the Protestant Reformation. In turn, the American work ethic served to drive **capitalism**. This work ethic explains why capitalism works so well in the U.S. Capitalism thrives off of people valuing work above other aspects of life.

Capitalist societies often distribute social resources (wealth, power, and prestige) based on two rules. First, the more important a job is to the daily functioning of society, the higher the pay. Second, relative to the demand for a job, the fewer the people who want or can do the job, the higher the pay. Accordingly, the hardest, most important jobs, such as a physician, pilot, or rocket scientist, will attract people with the most innate or learned skills, for which they will receive more wealth, power, and prestige. This model, essentially a functionalist argument, and an argument most Americans buy into, helps societies maximize collective efficiency by matching necessary roles to individual abilities.

Let's look at another example, that of garbage collector. Garbage management is vitally important to society. Imagine the impact if garbage collectors went on strike for a month or longer, leaving an entire city's trash to pile up outside of homes and businesses. The mounds of trash would eventually block the streets, preventing people from getting to work, and causing serous public health problems. But even though we need, and depend on, garbage collectors, and not many people want to serve as garbage collectors, the job of trash collection does not pay well. Nor does it receive high power or prestige. Why? Because the role of garbage collector does not require special skills.

Sometimes, though, a disconnect occurs between the demand for a job, the supply of available workers to do the job, and the job's corresponding salary, power, and prestige. This disconnect results in **class inequality**. Consider the job of professional basketball or football athletes. The athletes play a sport, which serves as entertainment to others. This job is not vital to societal functioning, nor is the demand for the job large. Further, there seems to be a large number of mostly young men, especially young African American men, who aspire to these jobs. So what accounts for the high wealth, power, and prestige that professional athletes enjoy? We could argue the difficulty of excelling in these jobs, and that even though many people want them, most people cannot perform them at a professional level. Before we conclude with that answer, let's look at another example.

The occupations of teaching and policing represent vital services to society. They are difficult to do, requiring special skills. The demand for these jobs seems stable, if not growing, and the supply low. Power accompanies both the roles of teacher and police officer in the form of authority and influence over students and residents. Yet both jobs receive relatively low pay and low social status. Why?

Applying conflict theory here may help. Conflict theory would say that the allocation of wealth, power, and prestige to jobs reflects the collective power of the people in those jobs, more so than the demand for and supply of the jobs. Take a look at the occupation of physician. We pay and treat doctors well, and we afford them a great deal of power in our lives. But that was not always true. At the end of the 19th century, we did not even call physicians "doctors." No standardized training or licensure requirements existed. While we treated them as professionals, back then physicians received much less prestige and pay than today.

The occupation of physician evolved through a complicated process loosely called **credentialism**. Practicing physicians, concerned about the status of their profession, came together to discuss professional interests, such as how to identify training requirements and how to validate the completion of appropriate training. Out of this, physicians created professional associations to self-enforce the new requirements and certifications and to represent their occupational interests. Their professional associations began lobbying state legislatures to protect citizens by licensing physicians to practice in each state. Once in place, the required degrees, certifications, and licenses served as gatekeepers for entry into the profession; thereby, maintaining or improving the quality of practitioners and legitimizing increases the cost of their services. As you might expect, credentialism often snowballs, resulting in higher and higher training requirements and salaries.

Recently, credentialism slid into other occupations and industries, such as social work, nursing, financial planning, and information technology to name a few. Just one look at the Sunday classifieds and you will see this pattern. More and more job ads request applicants to possess graduate degrees or some additional specific certifications or licenses. You can also observe credentialism by collecting business cards from friends and associates. You will notice that professionals from a wide range of occupations provide a laundry list of letters after their names, many of which contain no meaning to people outside their field. The letters stand for a specific degree, specific certification, license, or so on. In their profession, you must possess these various degrees, certifications, or licenses in order to "do business."

A more functionalist, and positive, interpretation of credentialism goes like this: In a rapidly changing society, new work roles emerge almost daily. Think of all the technological advancements that occurred in just the last 10 years alone. In this kind of environment, it becomes critical for people to continuously learn new skills. Regardless of how you see credentialism, if you thought you could stop learning new tricks after you got your college degree, you're wrong! **Life-long learning** is not just a buzzword, it's essential! So prepare to start collecting your own set of letters.

TOOLS

Applied sociology brings a unique set of tools to the field of work and occupations, a field that the business world often calls *organization development, human resource management* or *human relations*. The sociological perspective lends insight into job roles, team building, management styles, organizational change, conflict management, information diffusion, and strategic planning, to name a few. Applied sociology offers research tools to gather, analyze, and interpret data on, among others, job satisfaction, job turnover, organizational communication, and employee training and retraining. Let's take a look at an example.

A national grocery store, with thousands of stores, employs over ten thousand **white-collar** and **blue-collar** workers, such as, respectively, managers, accountants, marketers, and cashiers, butchers, bakers, fishmongers, and warehouse laborers. The owners recently installed computers in each section of the store and trained employees on how to use them to streamline their jobs and reduce paperwork. Soon afterwards blue-collar employees began talking about unionizing. The owners worry that a unionized staff will lead to strikes and higher labor costs, which would put the store out of business. Subsequently, they instruct upper management to discuss employee complaints with key blue-collar employees. From these discussions, upper management learns of widespread job dissatisfaction among employees. Employees fear that the new computers will replace them. Additionally, minority employees voiced discontent over being paid less than white men in the same jobs.

The owners' respond to the data with shock. They think the computer systems they installed across the stores make employees' jobs easier, and subsequently, more desirable to employees. They argue that regardless of gender, race, or ethnicity, employees receive the same pay for the same job. Overall, the data puzzles the owners. They decide to ask for help from outside the company, someone with a more objective position than upper management. They hire an applied sociologist as a consultant to make sense of the data and submit recommendations.

The sociologist analyzes the data using her qualitative analysis skills. She then utilizes her knowledge of social structure, social settings, and people's beliefs, attitudes, values, and behavior to interpret the patterns in the data. She sees alienation as the central problem that the employees experience. The data indicate that the blue-collar employees feel like the owners no longer value their work. Because their jobs no longer take as much thought or practice, they feel like the computers could easily replace them. The employees miss a sense of meaning and fulfillment from their jobs, more important to them than salary. They want jobs they enjoy, in which they feel secure and connected to their communities.

The sociologist explains that alienation is one of the negative consequences of **automation** and **deskilling**. Our advanced technology today enables us to use machines to perform more and more tasks that previously required human labor. In the process, we made jobs easier, requiring less skill. On the one hand, automation and deskilling reduce labor costs, which theoretically reduces the costs of goods and services to consumers, thereby improving a society's standard of living. On the other hand, the workers with automated and deskilled jobs become obsolete. Before automation, they contributed to the entire process of producing a product or providing a service. This encouraged a connection between the worker and his or her work and a sense of satisfaction from completing a task from start to finish. With automation, workers only contribute to a small part of a product or service.

For example, thanks to voice-activated answering systems, secretaries no longer need to answer caller's questions or know how or to whom to transfer calls. Managing calls may represent the only contact secretaries experienced with consumers of their company's product, providing a sense of connection to other employees and to the role their company plays in the larger community. Without that contact, the secretaries feel separated from the finished product of their labor, causing a sense of dissatisfaction and alienation. Similarly, cashiers' jobs now mirror assembly-line work where they simply scan each product over the censor, administer payments, and dispense receipts to customers. They no longer need to know the product, how much it costs, or even how to return correct change. Many stores today provide self-checkout lanes where consumers scan their products and pay with their debit or credit card -- a cashier is unnecessary.

With regard to the employee's perception of gender and racial inequality, the sociologist looks at the pay data on all the employees. In fact, the company pays employees in the same job more or less the same, with fluctuations due only to seniority in the company. However, **occupational segregation**, the sociologist points out, explains this finding. The company employs mostly white men in management, butcher, and fishmonger positions. While, women mostly occupy the cashier and secretary positions, men of color occupy most of the warehouse laborer positions. So the sociologist finds it no surprise that wages for identical jobs do not differ by gender or race when the job occupants themselves do not vary by gender or race.

The sociologist explains the concept of **comparable worth** to the owners. Positions that differ in title or job description but which require similar skill and responsibility should pay similarly. For example, the positions of warehouse foreman and the head cashier require similar skill and responsibility levels, and should subsequently earn the same pay scale. But the pay data shows that the grocery pays the warehouse foreman position, which only men occupy, more than the head cashier position, occupied by only women. Similarly, the warehouse laborer position and the butcher and fishmonger positions require similar skill and responsibility levels. The warehouse laborer positions, occupied by mostly men of color, pay less than the butcher and fishmonger positions, occupied by mostly white men.

The sociologist offers the following recommendations to the owners. First, they should offer **profit sharing** to all employees. This will provide an incentive for employees to work together to advance the grocery store, and an incentive for owners to connect with employees. Second, the owners should provide a forum for employees at all levels to share their ideas and listen to others' ideas on how to improve the grocery store. This forum will encourage all employees to participate in the decision making process, attain their buy-in on new programs and directions, and reinforce their role in the company. Third, the owners should change their pay scales to reflect comparable worth. Fourth, to limit occupational segregation, cultivate respect, and strengthen teamwork, the owners should **cross-train** employees and regularly rotate employees into different roles within the stores.

TASKS

1. While watching the news on TV with your parents, you hear a report from the federal Bureau of Labor Statistics on minority job applicants. Over the last decade, the number of Latino and African Americans applying for and receiving entry-level professional positions rose dramatically. But the number receiving subsequent promotions within the same company declined. This report confuses your parents who don't understand the contradicting trends. As a college student, with a sociology course under your belt, use the terms *affirmative action, token,* and *glass ceiling* to explain how and why these trends occur.

2. You work as a sociology university professor. A local union invites you to participate in a town meeting to raise awareness on the national trend of employers hiring more temporary and part-time workers. In particular, they would like you to present sociological information on why this trend occurs, how it affects temporary, permanent, part-time and full-time employees, and the immediate and future consequences of this trend for communities. Economists, psychologists, political scientists, and cultural geographers will participate in the town meeting as well. To prepare for this meeting, identify and outline the sociological issues and explanations you plan to present. Draw on the example of college courses being offered on the Internet to illustrate the sociological issues and explanations.

Chapter 8. Marriage and Family

QUICK START

In this chapter, you will learn:

- The functions and dysfunctions of the institution of marriage and family.
- Sociological explanations of marriage and family.
- How the institution of marriage and family influences, and is influenced by, other social institutions.
- How to assess the roles and distribution of resources within a family.

TERMS

Dysfunctions	Disruptive consequences of patterns of attitudes, beliefs, values, and behavior on societal operations.
Economies of Scale	The reduced costs of living or operation that occur when two or more households or organizations merge.
Family	The most significant primary group involving the relationships between children and parents.
Femininity	The beliefs, attitudes, values, and behaviors that people engage in to signify womanhood.
Functions of Family	Create and sustain groups of significant others; socialize members; maintain members' physiological, economic, psychological, material, ideological, and sociological needs; and transfer of social status.
Functions of Marriage	Designates appropriate sexual relations between adults, regulates sexual activity between partners, and provides a context for reproduction and parenting.
Identity	More than personality, this is who we think we are. Our notion of our self, which is constantly a work in progress.
Marriage	A significant primary group involving the relationships between romantic partners.
Masculinity	The beliefs, attitudes, values, and behaviors that people engage in to signify manhood.
Patriarchy	A way of organizing groups or societies to favor men over women in all social institutions, such as in families, laws, jobs, education, etc.
Pink Collar Jobs	Occupations associated with women, such as washing, sewing, cleaning, typing, teaching, and nursing, usually with poor pay.
Second Shift	The work that begins when a woman, who works outside the home for pay, returns home and provides all, or most, of the domestic labor in the home such as cooking, cleaning, and child care.
Sexist	Beliefs and behavior that discriminate, oppress, and alienate people because they are women.
Significant Others	Members of our primary groups with whom we hold intimate relationships, such as parents, siblings, best friends, partners, etc.

TOPICS

Most of us take **marriage** and **family** for granted. We take for granted our parents, brothers, and sisters, and our extended families of cousins, uncles, grandparents, and so on. We think our families "just happen," requiring little from us to maintain them.

Some families begin with a marriage, some not. Some end in divorce, some not. Just what is a family anyway? Is it people joined by marriage, by blood, by law? Is that all? Is someone that you treat "like a sister" a family member? Can we create families by collectively defining them that way? Sure we can, and we do it all the time. We bet most of you have "uncles" or "aunts" that are your parents' close friends, but not their brothers or sisters. Nevertheless, these people may play as big a part in your life as your parent's brothers and sisters.

If what we call things shapes how we behave, then certainly the use of the concept family shapes human action. For example, if a person defined as a family member asked to stay at your house or borrow your car, most of us would relatively quickly respond "yes" (granted, we would place restrictions on this arrangement for some family members!). If a stranger or an acquaintance from work or school asked for the same assistance, most of us would regretfully decline.

Most people do not think critically about marriage and family in our society, let alone our own marriages and families. Sociologists do. From a sociological viewpoint, marriage and family are institutions. Institutions emerge because they serve some purpose in the larger society. The institution of marriage and family exists to create and sustain groups of **significant others**.

What are other specific **functions of family**? We look to family for maintenance of our physiological, economic, psychological, and sociological needs. It's rare to find another social institution in which we expect people to love and care for us just because we were born! This function of providing for basic human needs occurs in a complex variety of family structures -- single parent families, two-parent families, gay families, and stepfamilies. Across all these types, families provide a host of material and ideological services to its members. Families are the main place of socialization, even with today's media presence in our lives. Families provide for members' economic needs such as shelter, food and clothing, and members' emotional needs such as love, friendship, and nurturing. Further, families transfer social status, values, beliefs, religion, recreation, education, and technology to their members. Families connect and bond individuals, providing a sense of belonging and preventing sexual relations between members.

Speaking of sexual relations, what are the **functions of marriage**? Well, even with the sexual revolution of the last 40 years, marital status still regulates sexual activity. Many people still frown on sexual relations outside of marriage. Many people still closely link marriage and reproduction. People still prefer for themselves and for others to enjoy the "benefit of marriage" before parenthood. So much so that expecting a child induces marriage for many people. Further, marital status still designates appropriate sexual or romantic partners. Married people are supposed to be "off the market"; hence the ritual of wearing wedding bands to indicate unavailability to others. Marital affairs still represent grounds for a divorce in most states.

Society depends on the institution of marriage and family, as well as the other social institutions, in order to run smoothly. To run smoothly, each institution depends on all the other institutions. Institutions such as marriage and family, education, the economy, and religion are interdependent parts of our society. For example, without our norms on how to conduct business (the institution of the economy), working parents (the institution of marriage and family) could not purchase child care. Similarly, other institutions depend on the institution of marriage and family. For example, unless teachers can assume that most children receive help, encouragement, and guidance with their school work from their parents or caregivers (the institution of marriage and family), it would be difficult, or impossible, to educate the country's youth using large public bureaucracies (the institution of education)

Marriage and family, like any other institution, also serves **dysfunctions** to society and other institutions. For example, theoretically, a high rate of divorce increases society's living expenses, because fewer people benefit from **economies of scale**. If people respond to increased living expenses by working more, they may enjoy less time for religion, education, recreation, and so on. A high divorce rate might also lead to an increased crime rate by not "settling down" large numbers of young men, the group that commits most crime. Over time, a high rate of divorce could result in a declining family size, which would dramatically affect a society's production and distribution of goods and services.

TOOLS

Looking at marriage and family as an institution can offer important insight into understanding the world we live in. How can we apply sociology to understand and solve problems concerning marriage and family? Well, this is a great place to use the theoretical tools we outlined in Chapter 2. Picking up on our conversation above, how would each theory explain the high rate of divorce in the U.S.? Let's begin with functionalism.

Functionalism

Functionalism would explain a high divorce rate by pointing out that the functions of marriage have changed as society has changed. The changes in technology, which fostered everything from birth control pills to genetic cloning, influenced the reproductive function of marriage. The economic transition from the farm to the factory changed the everyday structure of marriage. Let's take a closer look.

Today, men and women possess more control over their reproductive life. Increased reproductive control leads to smaller family sizes. Smaller families make divorce more feasible. As thousands of people show us, men and women with one or two children can, though not easily, live the life of a single parent. In contrast, men or women with six or seven children would experience great difficulty parenting these children alone, which may lead them to stay in unhealthy or unhappy marriages.

With regard to economics, in the past, families lived and worked together on farms and in small businesses, often operated out of their homes. This economic structure encouraged couples to stay married as their lives were too intertwined to separate. With the change to families working apart, away from the home, in different buildings and industries, married couples' lives became less closely linked. This makes it possible for couples to truly separate and live independent lives. Now we are seeing more people telecommuting to work, allowing them to work from home, and possibly near their spouse or partner. How will this drive the divorce rate? It could simplify people's lives, which may afford them more time and energy to invest in their marriage, resulting in a reduced divorce rate. On the other hand, the increased daily interaction between couples may lead to irritation and friction, resulting in an increased divorce rate.

Conflict Theory

To explain a high divorce rate, among other social forces, conflict theory would point to changes in the **patriarchy** of our society and social institutions. Men's hold on power and wealth is loosening. In the past, married men nearly ruled over their wives and children. Husbands sat at the head of the household, giving the final word on any decisions concerning the family, and no one -- including the husband, wife, children, extended family, friends, and community -- disputed it. Many women worked exclusively in the home, without pay. Those that worked outside worked in **pink-collar** occupations such as washing, sewing, cleaning, typing, teaching, and nursing, which paid much less then men's occupations. These women came home to a **second shift**, providing all the domestic labor in the home such as cooking, cleaning, and child care. Men also held political power in society, which allowed them to write laws declaring it impossible to charge a husband with raping his wife and impossible for women to own property or to vote. In every aspect of life, women depended on men, making it difficult to seek divorce.

Over the 20th century, the status of women in society changed. Women today seek and gain positions of equal political, social, and economic power to men. Women vote, own property, and work in just about any industry -- even broadcasting professional sports from men's locker rooms. While many married women still partially depend on their husbands for economic resources, most women work outside the home for pay. Within the home, many married working mothers demand that their husbands help with domestic labor such as shopping, cooking, laundry, and child care. Increased political, social, and economic power allows women greater freedom to end unhealthy and unhappy marriages. Equal contact with the world outside the home may also provide women with the same access to other sexual partners, which may result in more men seeking divorces than in the past.

Interactionism

Interactionism could provide many different explanations for a high divorce rate. Perhaps the simplest explanation would be that over the last 40-50 years, as a result of the changes described above under functionalism and conflict theory, our expectations of marriage changed. Many people no longer expect a marriage to stand the test of time. They may hope for and promise a lifelong marriage on their wedding day, but the option of divorce lingers nonetheless. Because collectively we expect less of marriage, people's perceptions of divorce changed. Divorce used to be taboo. Families, friends, and communities would label and stigmatize divorced men and women, especially divorced women, known as divorcees. Removing stigma makes it easier for people to divorce.

We could also attribute a high divorce rate to changes in **femininity** and **masculinity**. In the past, men and women valued women who acted fragile and submissive. Women, dependent on men and marriage, learned to act this way. We came to define this passive behavior as feminine. Women valued themselves and other women for acting feminine. But femininity has and is changing. As we describe above, more and more women seek and demand equal power and value in their homes and society. This helps women to value themselves and to construct strong personal **identities**. For example, more women today maintain their own identity when they marry by keeping their last name, and many extend their last name to their children. Because women think more highly of themselves, fewer may tolerate **sexist** husbands, leading to higher divorce rates.

Similarly, masculinity has and is changing. We used to value men for providing, protecting, and leading. While we still value these behaviors, we also now value men for their attentiveness, caring, and understanding. More and more men support gender equity and more and more women expect men to support gender equity. As a result of changes in masculinity and femininity, men and women became more responsible for themselves and more share responsibility for children. For men, the diminishing dependence of wives for sociological, psychological, political, and economic resources on their husbands lifts a past barrier to seeking divorce: guilt and responsibility. For women, knowing that men can and, many will, care for their children lifts a past barrier to seeking divorce: absent fatherhood.

TASKS

1. Several CEOs of local businesses threaten to retract the educational grants they awarded to a local high school if the school administrators do not remove a sociology text from the library. The text includes material on "alternative families." As one of the school's administrators, build an argument for students learning about alternative families, and for keeping the book in the library. You will present this argument to the local media and to the CEOs.

2. Use the sociology of marriage and family to reflect on your family system and map out the social forces and functions that impact on it.

   a. Create a cross impact assessment on your family. Write in the table below all the demands placed on your family from the outside -- the external forces from other institutions. Also consider the internal and external forces and expectations demanded by your family itself.

	Economic	Religion	Government	Community	Health	Education	Family
Family Member							

b. Consider your family's current division of labor. Write in the table below the person in your family with key responsibility for each of the demands indicated in the table above.

	Economic	Religion	Government	Community	Health	Education	Family
Family Member							

c. Review the relative distribution of tasks in your family. Do you detect inequities? Conflicting roles? Or other pressures in the family system?

d. Brainstorm with family members. Are there alternative ways that this work may be performed and/or distributed?

QUICK START

In this chapter, you will learn:

- The functions of the institution of education.
- The importance of life long learning.
- To identify what information should be taught.
- To identify the best ways of teaching information.

TERMS

Education	A social institution with the function of passing on skills and abilities that prepare persons to be effective members of society.
Anticipatory Socialization	The process of preparing members of society to take on roles, beliefs, values, norms, and attitudes that they will need to perform in their lives.
Synchronous Education	An educational system in which students and educators exist in the same time and place, and students take a passive role in their education.
Asynchronous Education	An educational system in which students and educators do not exist in the same time and place, and students take an active role in their education.
Resocialization	The process of transforming a person or group's set of beliefs, values, norms, and attitudes, which results in new societal behavior and concept of self.
Take the Role of Other	The process through which people see the world through another person or another group's perspective or position.

TOPICS

We spend a large part of our lives in school. Sociologists call the complex system of transferring information **education**. Our institution of education represents a commitment of society's resources and energy to provide knowledge to its citizens. Education is one of several institutions, such as economy, work, marriage and family, that serve various functions in society.

With regard to the institution of education, a sociologist might ask, what manifest functions does the education institution need to perform? An obvious function is to transmit basic educational information to students about reading, writing, math, information technology, and so on. But at all levels -- day care, child development centers, preschool, kindergarten, elementary school, middle school, high school, and post secondary schooling -- society benefits from several other important latent functions of education. Let's take a look.

To begin, the institution of education in the U.S. teaches students how to "do" gender appropriately roles. In day care, child development centers, preschool, kindergarten, and elementary school, children learn what games are appropriate for girls and boys. For example, in school, we encourage girls to play with dolls, to learn how to play individually, or in small groups, and how to **take the role of other**, which teaches empathy. We encourage boys to play complex games with rules, large opposing teams, and leaders. This teaches them competition, team membership, and leadership, preparing them for employment. At higher levels of education, girls and boys learn what behaviors are most valued for women and which are most valued for men. For example, the most popular boys are usually those in high profile positions on sports teams, and the most popular girls are often the cheerleaders, homecoming queens, and so on. Given these patterns, it is not surprising that after 12 or more years spent in mandatory educational institutions, girls and boys develop different personal skills and career goals.

The education institution also serves a custodial function. While young people, age 5-18, are in school, parents do not have to watch or care for them during those hours. This allows parents to more effectively and efficiently work in and outside of the home each day. Imagine working as a lawyer, doctor, teacher, police officer, bus driver, or janitor and supervising and caring for your children at the same time. It would be very difficult, and would lead to you produce less at your job. Or it might lead many people to leave their children unsupervised all day, which would likely result in much more community deviance and crime among youth than we experience today. Large numbers of supervised youth might also lead to higher numbers of them having children, marrying, and entering the job market to provide for their new families. Bottom line, without the institution of education, our institutions of economy, criminal justice, and marriage and family would change dramatically.

Our institution of education also provides **anticipatory socialization**, one of the most important functions of education, besides transmitting basic information. Let's think about this. If we anticipate something, we prepare for it. Socialization is the passing on of norms, behaviors, beliefs, attitudes, values, and skills from one group to another. Socialization enables people to interact with others in varying social situations. During 12 years or more of mandatory education, our schools teach American youth social skills that they will need in order to function in our society as adults. Schools teach students about the norms, beliefs, attitudes, and values of our society, and how to interact with people of similar and different demographics. Schools also teach students how to understand, follow, and comply with rules -- critical skills for their future roles as employees, parents, and citizens.

The anticipatory socialization function of education causes us to ask two things. First, what must citizens learn in order to interact effectively with others? Second, how should we transfer a system of appropriate, culturally relevant learning to citizens? Let's take these in turn.

We probably all agree that people need to be able to write, read, and perform simple mathematical calculations like adding, subtracting, and dividing. What else do people need to know in order to be productive participants in our society? When societies change slowly, what people need to know changes slowly. When societies change fast, as ours does, what people need to know changes fast. Rapidly changing societies "change the target." As the velocity of social change increases, the tension between what needs to be taught and the social structures that teach it increases. Hence, those looking for the "core" curriculum of our educational system will quickly realize that the core is relative to the culture in which the school exists. Meanwhile, the diversity among learners varies, too, making a society's educational challenge even greater. In a rapidly changing society, learning must be life-long. Furthermore, in an ever-changing society, it becomes increasing hard to answer the question, "Education for what?" Should education institutions target the present, the future, both or neither? Our chapter on the future (Chapter 14 Social Change and the Future) addresses some of these thoughts.

The second issue surrounding education is how should it be done? Most industrial societies construct large bureaucracies of **synchronous education** to produce fairly routine educational outcomes. After assembling in a common place at a common time, sets of teachers transmit a common set of information and skills to a set of students. The rigidity of some of these systems likens them to stagnant, learning factories. For example, your experience in the education system probably did not differ significantly from your parents' experience. In contrast, recent changes in information technology supply users with an alternative, **asynchronous education**. For example, on the Internet, users actively engage information that interests them, and do so on their own schedule, not the system's schedule. Life-long learning requires asynchronous education, and most institutions today -- business, government, and health care -- need to promote life-long learning in order to keep workers, citizens, and consumers trained and informed in a rapidly changing society.

The asynchronous nature of the Internet questions some long-standing norms, roles, and statuses -- namely, that of teacher and student. It questions the teacher as a one-way transmitter of information to a passive assembly line of students. In contrast, the Internet offers wide and immediate access to information. The Internet tries to keep pace with the dynamic society in which students live. Because of the bureaucracy of our educational system, teachers rarely keep pace with social change. As countries transition from industrial to information-based economies, teaching and learning become simultaneous, and citizens become omnipresent. Consequently, the need for constant **resocialization** becomes essential. This transition challenges our educational system, our teachers, and our learners to develop the skills needed to be both learner and educator.

TOOLS

So how does applied sociology figure into the challenges facing our educational system? Well, not only will the application of a sociological view indicate "where" society is going, but a sociological view will also guide how to obtain, create, package, and deliver information for use in our society. Sociologists and other social scientists projected for several decades that our economy would move toward information and services and away from material production. In the next decade, we will likely see increased access to information technology among senior citizens, the economically disadvantaged and other groups who currently use computers or the

Internet in small numbers. Similarly, more people will work outside of their homes via telecommuting. More institutions will deliver information and services via inter- and intra-nets, such as health care, education, politics, religion, and criminal justice. For example, according to Mary Beth Susman, the former executive director of the Kentucky Commonwealth Virtual University, citizens of Kentucky can currently apply to all state colleges, public and private, via one electronic application. Then, once accepted, they use one interface to register and take courses at any combination of schools. Via the Internet, they can access libraries at every school in Kentucky, form on-line learning groups with other students, and complete their degree without ever stepping foot on a brick-and-mortar campus.

The sociological perspective on education will also help make sense of educational processes in other aspects of society. For example, as we mentioned in an earlier chapter, businesses today, especially information technology companies, suffer from high job turnover. Many companies ask, why is this happening? What can we do about it? As a sociologist, you could see turnover as a failure of the organization to properly socialize its employees. Through socialization, we not only learn how to act in-groups, but also develop attachments to groups. People who change their jobs lack attachment to the organization employing them. Businesses may need to provide more opportunities for employees at all levels to interact with one another in and outside of work. For example, they may want to sponsor employee athletic teams, continuing education and training programs, ceremonies acknowledging the achievements of employees, improved day care and health care programs, and community education. In doing so, employees become connected to their employer in many ways, resulting in increased loyalty and decreased turnover.

A more obvious use of applied sociology in the field of education is to help determine what people need to know and how they should come to know. For example, a state Board of Education may hire you to determine what information and skills high school students need to learn and the best way to teach them. To answer the first question, you might conduct a needs assessment among the state's major employers on the skills they need now and will need in the next five years (see "What Employers Want" in the *Job Outlook 2002* by the National Association of Colleges & Employers[7]). You might combine these results with the sociological predictions for an information economy and advise the school board to center the high school curriculum around communication skills, computer skills, and problem solving.

To answer the second question of how to teach high school students, you might draw on the sociology of education literature and advise them to organize the children into learning groups, rather than into individual desks. The students would work together in their learning group on in-class assignments. Schools would shuffle the groups a few times a year. With this organization, students would learn how to work in-groups, how to handle conflict, how to lead, how to follow, and how to communicate effectively with others. This organization would also cultivate respect and acceptance of diversity and reinforce social bonds to others, which would lessen social problems in the future.

---

[7] Also available at http://www.uncw.edu/people/pricej/teaching/employerwants.htm

You could also answer the above questions by employing an affinity diagram. For purposes of this example, let's assume that you're talking to a group of parents and educators at a local private school. You've been invited to help them "get a handle" on the best way(s) to teach their children. Of course, you can use this tool in a variety of situations, outside of education!

Affinity Diagram[8]
1. You will need a group of people from an organization or community
2. Define the problem. What do you want to know? In this case: "What are some ways to best teach our high school children?"
Distribute 4-5 "post-its" to each person in the room. Make sure that everyone has something to write with -- preferably a pen or pencil.
3. Ask all persons to individually write down as many "best ways" to teach our high school children.
4. Ask participants to stick the post-its on a board, wall, or black board. No particular order is required.
5. After everyone completes the above task, ask them to come stand next to the posted post-its. Then ask them as a group to sort out the ideas into groups -- meaningful categories by physically moving them around into similar bunches. This should produce discussion and sometimes a little tension as people move post-its back and forth. Often 1-2 individuals will begin to lead this process.
6. When the participants form categories, ask 1-2 people to create a heading for each category of items. Provide them with another post-it to serve as the category heading.
7. Ask everyone to take their seats. Read the category headings and some sample post-its from each.
8. Verify that everyone agrees with the categories and content. Based on feedback here, consider reorganizing the categories.
9. Now proceed with your discussion addressing each of these categories, type up the categories and headings, and distribute them by email, or schedule another meeting to discuss the findings.

TASKS

Create an outline for an educational workshop on energy conservation to a selected group.

1. Do a needs assessment -- decide the "who, when, and what?" of the training.
    - Determine the target population for the workshop. Who should receive the training or the education? Define the limits of this group.

    - Now identify a person or group as your client (for whom you will provide training). With them, define the length of the training.

2. Determine the needs of those in the group. Consider using "key informant interviews." These interviews focus on the persons perceived to have the major impact in deciding what the training or workshop ought to be.

3. Interview the key informants on what they think the training should cover.

---

[8] Adapted from Brassard, M. and Ritter, D. (Eds.) 1994. *The Memory Jogger II*. GOAL/QPC: Methuen, Massachusetts. Pages 12-16.

4. Create a training outline, including the mission, goals, and objectives of the training.

- State the purpose or mission of the training. Answer this question: "The purpose of this training is to…"
- Then list the 3-4 main things that need to be accomplished in the workshop, these are the training goals, such as:

  Goal 1:
  Goal 2:

- Identify the objectives of each goal. In other words, for each of the goals, decide what each person who takes your workshop should be able to do, know, or feel after the training. For each goal you should have at least two objectives. To write an objective, finish this statement: "When this workshop is over participants should be able to know, do or feel _____," such as:

  Goal 1:

  > Objective 1:
  > Objective 2:

5. Check with your training client (the person in 1b).

- Give a copy of your workshop outline (typed) to your training client.

- Ask for feedback. That is, find out from your client if the workshop outline appears to meet their needs, and if you need to rework any aspect of your outline.

6. Produce the final workshop outline.

Chapter 10. Religion

QUICK START

In this chapter, you will learn:

- The social foundations of religions.
- To identify religious diversity.
- The role of religion in current and past societies.
- How to measure religious affiliation and religiosity .

TERMS

Agnostic	A person who neither believes or disbelieves in a supreme being(s).
Church	Types of religions, such as Protestantism, Catholicism, Islam, and Judaism.
Cults	Like a sect, but leaders and members go a step further by trying to change, through peaceful or violent means, the mainstream society.
Denomination	Types of religions, such as Baptists, Methodists, Episcopalians, Lutherans within Protestantism.
False Consciousness	The condition in which people who are being exploited, oppressed, or alienated come to legitimize their subordinated position and no longer see themselves as disadvantaged.
Macro	Using a sociological perspective to understand large-scale, national, and international issues.
Meso	Using a sociological perspective to understand large group and local issues.
Micro	Using a sociological perspective to understand individual and small group issues.
Religion	Any set of institutionalized beliefs and practices that help people understand the unknown.
Religiosity	The importance of religion in peoples' lives.
Religious Affiliation	The religious group that a person belongs to, such as Buddhist, Jewish, Baptist, Methodist, Catholic, Islamic, atheist, or another church, denomination, sect, or cult.
Religious Beliefs	Ideas about supreme beings, the truth of which cannot be proved or disproved.
Religious Rituals	Behaviors that represent religious beliefs such as lighting candles, praying, chanting, etc.
Sacred	The aspects of our selves, lives, and communities that do pertain to religious beliefs or behaviors.
Sect	A religion that professes to be uniquely correct and which separates members from non-members both physically and socially.
Secular	The aspects of our selves, lives and communities that do not pertain to religious beliefs or behaviors.
Secularization	The decline of religious importance in daily life.

TOPICS

The sociology of **religion** is not about verifying the presence or absence of a supreme being or a religious explanation of life. Rather, sociologists use the tools discussed throughout this book to "make sense" of how people organize and construct religion and how religion influences social life. In this way, religion serves as a social institution. Many people, though, find religion a sensitive topic for discussion.

Religion yields a powerful force in human societies. Religion provides people with ways of seeing the **secular** life world in light of the **sacred**. Religion explains the unexplainable and ties humans together to greater beings. In doing so, religion imparts sacred support for patterns of human action. For example, across cultures and history, people marry in a church, surrounded by friends and family, listening to the words and actions of a religious leader. Many would not consider getting married any other way. Why? Because of **religious beliefs** and **religious rituals**. Similarly, religious beliefs and rituals largely inform and shape our practices surrounding death and dying. Catholics hold wakes, Jews sit shivah. Some religions advocate the preparation of corpses for visitation by the living, others do not. Kneeling, touching, and talking to a corpse might seem bizarre without an understanding of the religious context.

From a sociological perspective, in order to understand social behavior, we must understand religious organizations and their impact on society. How can you use sociology, a science, to explain religion, often considered the opposite of science? Let's take a look. At their root, science and religion are similar social institutions. They both try to explain the unknown. But, they use different methods to develop their explanations. Science bases explanations on observations of the world, whereas religion bases explanations on faith. Objectively, at face value, many people see scientific explanations as more sound or rational. As science expanded over the last century, we witnessed the widespread **secularization** of societies. However, science can't tell us what we should or should not do. Those are moral and ethical decisions, often informed by religious beliefs.

Most people think of well-known **churches** and **denominations**, such as Catholicism, Judaism, Hinduism, and Protestantism (Episcopal, Baptist, Methodist, etc.) when they think of religion. Any organized set of beliefs and rituals that give meaning and understanding to people's lives can represent a religion. In this sense, for some people, music could be a religion, or art, science, meditation, exercise, and the list goes on. From a sociological perspective, groups of people build and change religions to make sense of their lives. Holding the same religious beliefs and practicing the same rituals binds people together and creates a shared reality and community.

To indicate community, most religions try to distinguish members from non-members, some more so than others. In fact, some, such as sects and cults, go so far as to create new societies. For example, **sects** such as the Amish live in communities separate from mainstream society. Sects and mainstream society co-exist peacefully. In contrast, **cults**, such as David Koresh's Branch Davidians in Waco Texas, use different methods to remove themselves permanently from mainstream society.

Religion also acts as a mechanism of social control. Many of our laws descend directly from Judeo-Christian beliefs and values. Less formally, expectations for individual and social behavior often come from religious beliefs. For example, our history of volunteering and giving charity to those in need flows from religious beliefs. Our religiously based customs are not always positive. Because of religious beliefs, many men expect women to defer to them, and many women prefer to defer to men. Similarly, many people used various religions for many years to legitimize slavery. So, you see, religion also contains the power to oppress, exploit, and alienate.

Sociologically, we could view religion from a **macro, meso,** and **micro** perspective. On a macro scale, religious belief systems are powerful and transcending. Across history, religious groups migrated globally, creating new nations (often through war), to spread or maintain their religion. In response to spiritual diversity, some societies hold on to a single religious order, sometimes integrating it with other political and social institutions. England, for a long time, embraced one dominant religion (the Church of England), which dictated political, legal, economic and social norms, even confirming royal leadership. In contrast, some societies encourage and allow for pluralistic, religious beliefs. For example, pilgrims left England in order to practice different religions, and helped to form their own country that protected religious freedom.

At a meso level, a local church, mosque, or synagogue often represents an integral part of the local social organization. As such, religion gets involved in all of the daily challenges of every aspect of social life. Looking back on Chapter 4 (Groups and Organizations), you could apply sociological tools to assist local churches or religious groups in everything from creating a unique religious identity, to organizing a religious school, to increasing a church's abilities to meet local social needs. Religious groups often find themselves in situations in which they must make many secular choices. You could use sociology to help them make these decisions. For example, churches today may want to offer (1) job training, (2) health care to the uninsured, and (3) adult education for people without high school degrees. However, they may not possess enough resources to support all of these options.

From a micro level, individuals internalize religious beliefs in combination with a variety of other social forces. Religious beliefs remain a primary aspect of a family's socialization of its children, and for some, a primary aspect of their child's education. Religious beliefs account for why people marry, how they raise children, how they deal with illness and death, how they interact with people from different walks of life, how they handle conflict, how they recreate, and what ideas and social behaviors they value and hold as true.

TOOLS

By now, you might be thinking, "Well this religion stuff is interesting, but what can I do with it? How I can use the sociology of religion to help solve social problems?" Let's walk through some applications. To begin, assume you work for the United Nations (U.N.). The U.N. develops programs to educate and distribute birth control methods in developing countries because reducing family size would dramatically improve the quality of life in these countries. Unfortunately, the program fails. Based on interview data, the U.N. learns that people reject birth control methods due to religious beliefs. Your boss asks a team of coworkers, including you, to explain why people choose to uphold their religious beliefs instead of improving their living situations. From taking a sociology course in college, you suggest that the team start with sociological theory as a tool to explain social behavior.

Using your functionalist theoretical tool, you argue that religious groups serve a purpose in the society and the culture in which they emerge. We can trace most religious beliefs and behavior back to some collective human need, such as the need for informal social control, the need for meaning and purpose in life, the need to connect people with similar beliefs, values, and behavior together and the need for people to want to belong to such groups. Without this connection and belonging, groups would likely disband.

Using conflict theory, you also explain that religion helps maintain inequality, such as sexism and patriarchy. For example, by men and women rejecting birth control, women of childbearing age often spend much of their life pregnant or caring for children, preventing them from performing other social roles such as community leadership positions that might result in access to and ownership of power and resources. As such, religion serves to maintain the status quo of male control. Additionally, religion promotes **false consciousness**. Religion promises nirvana if believers endure their current life. Belief in this nirvana, or heaven, encourages people to accept their plight in life rather than work to change it. Because of false consciousness, religious people in developing countries may strongly oppose birth control. Rather than seeing the chains that their religion places on them, women may see an opportunity to serve God and men. As Marx said, "Religion is the opium of the people."

Finally, using interactionist theory, you explain that groups of people socially construct religion. That is, people create religion based on the social context in which they live. If they live in a time and place in which their primary activities involve searching for food, preserving shelter, and protecting themselves and their family, people will likely look to simple beliefs and values to make sense of their hard living. In a context in which death and dying occur frequently, people may come to value birth as the only means of community survival. In a life with few pleasures, the joy of sex and of holding your own child may take on even more value. Finally, in a setting devoid of artificial technologies, people may find any device or method to prevent pregnancy highly suspect.

In addition to explaining religion, sociologists also measure religion. **Religious affiliation** and **religiosity** represent two of the main sociological measurements concerning religion. Let's take a look at how these two terms differ. Leaders of a large midwestern city would like to help the local homeless who live on the streets, especially during the cold weather months when sleeping outside often leads to death. They ask citizens to send in their ideas for solutions. One they receive says to allow the homeless to sleep in local grade school buildings and churches. These buildings remain primarily empty during evening hours. This suggestion would require no new building programs, which would increase city expenses and taxes. As a community relation's specialist for the city, you need to assess the level of support for a proposal among city residents.

You decide to conduct a telephone survey of residents. From your college sociology course, in the survey, you know that you need to ask residents about their religion. You suspect that members of some religious groups will provide more support than others for this proposal. What exactly should you ask about religion? First, you should identify respondents' religious affiliation, such as Buddhist, Jewish, Baptist, Methodist, Catholic, Islamic, atheist, or some other church, denomination, sect, or cult. No need to reinvent the wheel here, as valid and reliable questions to measure religious affiliation already exist. (With any project, whenever possible, you should use existing measures validated by previous research.) The religious affiliation questions below come from the General Social Survey, a nationally representative survey of the contiguous United States, conducted every two years. The GSS contains dozens of questions on religious affiliation and religiosity, as well as on hundreds of other topics.

- What is your religious affiliation? Is it Protestant, Catholic, Jewish, some other religion, or no religion? Buddhism, Hinduism, Other Eastern, Moslem/Islam, Orthodox-Christian, Christian, Native American, Inter-Nondenominational, Don't know, No answer

- If you are Protestant, what specific denomination are you, if any? Episcopal Church, Presbyterian, Lutheran, Methodist, or Baptist

- If Jewish, do you consider yourself Orthodox, Conservative, Reform, or none of these? Orthodox, Conservative, Reform, or None

- When it comes to your religious identity, would you say you are a Fundamentalist, Evangelical, Mainline, Liberal, or do none of these describe you? Fundamentalist, Evangelical, Mainline, Liberal, or None

Another way to think about religion is, "How religious are people?" "How important is religion in our lives?" These questions reflect religiosity, on which people vary dramatically. How should we measure this? Should we ask people how religious they are? Or should we ask them about their religious behaviors? (Behavioral questions often evoke more factual responses than attitude, belief, or self-identification questions.) Again, don't reinvent the wheel here. Adopt religiosity questions from any number of past or present research studies, such as the GSS. We adapted the religiosity questions below from the GSS.

- Do you attend church regularly? Yes or No

- Do you or anyone else in your family say grace or give thanks to God aloud before meals at home? Yes or No

- Do you participate in church activities? Yes or No

- Are you confident in the existence of a God? Yes or No

- Are church teachings important to you in making everyday decisions? Yes or No

Regardless of the tools sociologists use, our initial point stands: Sociologists focus on how humans construct religion and religious beliefs, not the "rightness" or "wrongness" of these beliefs. A person does not need to be atheist or **agnostic** in order to use sociological concepts and tools to understand the relationship between religion and society.

TASKS

1. Members of a national church disagree on the utilization of recent medical advances. Specifically, members disagree on using in-vitro fertilization, surrogate mothers, abortion, genetic cloning, organ harvesting, and methods to identify a baby's sex or presence of cancer genes. Some members threaten to split off into their own church. The national headquarters of the church hires you, a sociological consultant, to propose methods to resolve the division or to guide a healthy split. Use your knowledge of sociology of religion, and sociology in general, to outline some possible solutions.

2. A religious congregation wants to build an addition to its current place of worship. Use the tools in this chapter and others to outline the steps you would take in planning this expansion.

3. A nursing home or assisted living organization wants to discover how satisfied its residents are with living in their community and with the various programs offered such as food, recreation, health care, and transportation services. What data collection method(s) would you choose to answer these questions? Will you ask residents about the role of religion in their lives? Why or why not? If yes, what would you ask?

QUICK START

In this chapter, you will learn:

- The social causes of health and illness.
- The nature of health care delivery as a social system.
- Some models for thinking about health, wellness and illness.
- Some of the social outcomes of illness.

TERMS

Acute Illness	Severe, immediately life-threatening illnesses.
Chronic Illness	Prolonged illnesses that are not immediately life-threatening.
Comprehensive Health Care	Health care programs that cover health issues from head to toe.
Dominant Group	The group in a population that has more wealth, power, and prestige.
Emergent Care	Health care for illnesses that need immediate attention, such as heart attacks, strokes, and trauma injuries.
Fee-for-Service Health Insurance	Insurance that pays 80% of all covered health care expenses.
Health	An individual or group's biological and psychological fitness.
Health Care Access	The level of ease of entry into an available health care system.
Health Care System	A set of interdependent parts that provide health care resources.
Health Care Utilization	The rate of use of the system organized to provide health care in an identified target population.
Health Risk Behaviors	Human action or behavior that increases the likelihood that a person or group will acquire an adverse health condition.
Incidence	The number of times something happens each year.
Medicine	A social institution that provides health care to members of society.
Managed Care	Health care insurance that pays all covered health care expenses except for a per service co-pay.
Medicine	The institution the provides health care resources.
Medicaid	Subsidized health care for impoverished or disabled people.
Medicare	Subsidized health care for the elderly.
Medicalization	The use of medicine to understand and respond to social issues.
Minority Group	The group(s) in a population with the least wealth, power, and prestige.
POET	An acronym for analyzing a social problem in terms of population, organization, environment, and technology issues.
Prevalence	The number of times something happens across a lifetime.
Primary Care	Health care that focuses on health maintenance, illness prevention, and the diagnosis and treatment of common ailments.
Scapegoat	Blaming one group for a social problem caused by another group.
Socialized Medicine	Health care that is provided for all citizens and paid for out of taxes, like public education.
Urgent Care	Health care for illnesses that need immediate attention, such as broken bones and cuts requiring stitches.

TOPICS

Did you wake up with a cold this morning? Sneezing, coughing? The flu? What's your "mood" today? Happy? Sad? We understand all of these conditions in terms of **health** and wellness. While you no doubt think of health and wellness as something highly personal, sociology also sits at the core of our thinking here. Health is as much a social phenomenon as it is a biological one. In fact, health and illness are so important in our society that sociologists refer to the production, consumption, and delivery of health care resources as a social institution. Like other institutions, a society's characteristics influences the institution of **medicine** and the institution of medicine influences the social conditions of a society. Let's look at both of these influences.

To begin, social interaction shapes people's attitudes, values, and behaviors, leading to patterns of sickness or health. For example, people who highly value money may work excessively, which can lead to stress, in turn leading to various illnesses. Similarly, the followers of some religions, such as Jehovah's Witness and Christian Scientists, believe in using only natural remedies for poor health, and no invasive procedures. With **acute illnesses** or **chronic illnesses**, these behaviors could further impair health. Others, such as Mormons avoid caffeine, tobacco, and alcohol, which may result in improved health. Our attitudes, values, and behaviors also influence how we respond to illness in others. For example, our reference groups and norms influence how we react to AIDS.

Cultural patterns influence health and wellness. For example, wide regional patterns exist across the U.S. in **health risk behaviors** such as nutrition, exercise, and tobacco use. In the South, people frequently smoke, eat fried food, and avoid exercise. Subsequently, social scientists refer to this region as the "stroke belt," with a high **incidence** of diabetes, stroke, and heart disease. In other parts of the country, such as Western states, exercise holds a predominant position in the culture and social structure, leading to some of the lowest incidences of obesity.

Now let's look at the influence of the medical institution on society. To begin, in the U.S., the medical institution exercises unparalleled power and control over our lives. For example, think about how we treat physicians. We grant them vast authority, listening to their words of advice. Many people never question their physician's opinions or treatment plans, preferring a passive role in their health care. How do you act when you interact with a physician? Do you let them lead the conversation? Do you come prepared with a list of questions or symptoms? Many of us take a more active role in the "health" of our cars than we do in our own bodies. Do you refer to your physician as "Dr. X", while he or she calls you by your first name? How long do you usually sit in the waiting room and examining room before a physician sees you? Many of us wait over an hour, or even two hours. Would you wait that long for another professional such as a teacher, lawyer, architect, accountant, or information technologist? Further, we pay physicians handsomely and shower them with immense status. In other societies, physician's economic, political, and social characteristics match those of other professionals.

In recent years, the social institution of medicine spread into previously defined "non-medical" aspects of our lives. Sociologists refer to this spread as the **medicalization** of society. We now think of smoking tobacco, drinking alcohol, overeating, and gambling as addictions, diagnosed and treated by psychologists, psychiatrists, or other physicians. Young people tune in to radio and television programs daily to hear how their poor childhood relationships with their mothers or fathers caused their current academic, sexual, and economic problems. The medicalization of the above problems ignores their social causes instead reducing them to individual psychological or biological conditions. For example, regardless of parental mistakes, women living in a social context of patriarchy and sexism will likely experience repeated bad relationships with men. However, by medicalizing these problems, we eliminate our responsibility for messy social problems, and avoid making the difficult social and cultural changes necessary to solve the problem. Instead, we opt for a neat pill or therapy to "treat" the problem.

The social institution of medicine also holds the power to label people as "sick" and worthy of our empathy or pity. For example, chronic fatigue syndrome (CFS), also known as Epstein Barr syndrome, is a relatively new diagnosis characterized by extreme tiredness. Until recently, health care providers did not diagnose or treat people with these symptoms, and the medical institution, including pharmaceutical companies, distributed little information about these symptoms in medical communities or in the mass media. Subsequently, we stigmatized people who complained of being too tired to work or care for their families as being lazy. Now we see CFS as a known and treatable illness. Today, many people suffering from extreme tiredness receive help from their families, friends, and the medical community. Further, we now grant people suffering from CFS special legal protection in their jobs and various social assistance programs such as disability. Similar processes occurred with other illnesses such as diabetes, Tourette's syndrome, Alzheimer's, AIDS, carpal tunnel syndrome, and attention deficit disorder.

The labeling process above also works in a reverse pattern. The medical community no longer identifies homosexuality as a medical condition, nor does it think homosexuality should be or is medically treatable. However, many Americans, both heterosexual and homosexual, still consider homosexuality a medical condition, caused by physiological or psychological misfortune. Some homosexuals still seek treatment such as "conversion" therapy. Many heterosexuals pity homosexuals as "defective," as less than fully human, and therefore, not worthy of civil rights, social privileges, or socio-economic opportunities (much like the historic discriminatory experiences of African Americans).

Sociologists call this reaction **scapegoating**. People in power, in this case heterosexuals, tend to place blame for problems on others, because that way they can ignore problem at their roots, which usually lies with the dominant group's attitudes, beliefs, values, and behavior. Scapegoating allows a **dominant group** to feel no guilt or responsibility and instead shames the **minority group**. The root of the problem, in this case, is heterosexism and homophobia, not homosexuality. Changing heterosexism and homophobia requires large scale change in our social structure and social institutions. It also requires heterosexuals, and homosexuals who internalized homophobia, to change their behavior, values, and attitudes.

TOOLS

In trying to understand the influence of and on the social institution of medicine, sociologists draw on theoretical, conceptual, and measurement tools. Let's take a look.

What happens when you get sick? Well, Americans often just wait for the illness to pass. If it doesn't, then we "go to the doctor." Where do we get this idea? It is a culturally relative decision to rely on a health care system. Here's a great place to use our functionalist tools to make sense of the institution of medicine. We learn that "when we get sick, then, we go to the doctor." This taken for granted approach reflects an acute care notion of disease and wellness. Culturally, we learn "If it isn't broken, don't fix it! Wait until you're sick, then worry about it!" Why do we use a reactive approach to illness, rather than a proactive, preventive approach to wellness? Now with modern advancements in health, wouldn't it make more sense to do everything possible to prevent illness?

At the macro level, using functionalism, we see that a social system or institution emerges because it serves a purpose. Social institutions evolve and change in response to the social setting in which they exist. Our **health care system** results from four forces: (1) population characteristics such as size and life span; (2) the organization of our health care system such as **fee-for-service**, **managed care**, or **socialized medicine**; (3) environmental factors such as water quality, waste disposal, and climate changes; and (4) technology such as the development of scientifically based drugs and equipment to cure illness and prolong life. In short, we can use the concept **POET** (population, organization, environment, and technology) to understand the development of the institution of medicine, as well as other institutions.

Using POET, we see that prior to the 20th century, life expectancy was short, water and waste disposal systems were poor, infectious illness was common, and disease immunization was only an idea, not a practice. At the beginning of the century, parents frequently buried their children due to a high infant mortality rate. Thanks to advancements in water, waste disposal, and health care systems, by the mid 20th century, infant mortality declined and lifestyle factors such as diet, sleep and exercise improved. By the end of the 20th century, life expectancy doubled, with the 85 and older age category showing one of the highest growth rates. Noting these changes in population, organization, environment, and technology, what kind of changes should we expect in the institution of medicine in the 21st century? We could write a whole book on this topic. In short, with larger numbers of older people in our society, we should expect the **prevalence** of chronic illnesses like Alzheimer's, diabetes, and arthritis and acute illnesses such as cancer, stroke and heart disease to increase, requiring us to either produce more health care services or change the way we distribute them. These changes would result in subsequent change in our institutions of the economy, politics. marriage, family, education, housing, and so on.

Moving to the meso level, using conflict theory, we see how the diversity factors unfold. **Health care access** and **health care utilization** are closely linked to inequality. In the U.S., gender, age, race, and social class all limit people's access to health care resources. People primarily access health care resources in the U.S. through fee-for-service health insurance provided by an employer. Most employers fail to offer health insurance to their part-time employees. Many small businesses fail to offer health insurance or other benefits to their full-time employees. Who falls into one of these non-insured groups? From Chapter 6 (Diversity and Inequality), we know that this group contains disproportionately high numbers of women, the young, non-whites, and people of the working class. Many of the uninsured do not earn enough to afford to purchase medical care privately, but they earn too much to qualify for **Medicaid**.

In contrast, in other developed countries, all people, regardless of gender, age, race, or class, benefit from access to health care resources via a socialized health care system. Countries like Canada and England consider health care a human right, much like Americans consider freedom a human right, and distribute health care to everyone, according to need, via government programs. The U.S. offers free or discounted health care to some citizens. Our tax dollars provide **Medicare** for those over 65, Medicaid for those in poverty, health care for military employees and their dependents, and a growing number of health care programs for children. While these programs mirror socialized health care, Americans do not refer to them as such.

Moreover, gender, age, race, and social class culturally limit people's utilization of health care resources. For example, men less often seek medical treatment than women. The desire to be and appear masculine makes men act tough. Subsequently, higher numbers of men die of heart attacks than women because men ignore the warning signs. Similarly, delaying medical attention, men are often diagnosed in the later stages of cancer, with less effective treatments. Gender influences women's health care utilization too. More women than men receive a psychological diagnosis and treatment for physical problems, with physicians often telling women to "rest," implying that the problem lies in women's heads. When men seek medical care, physicians take their problems seriously, ordering a host of medical testing.

Additionally, social class and race influence health care utilization. For example, many obese working class Americans do not realize that they have diabetes, and subsequently do not seek medical treatment. Similarly, African American women are less likely, than other women, to perform monthly breast exams, or to seek treatment for suspicious lumps in the breasts. If they do seek a doctor, they often choose non-medical treatments such as praying and folk medicine. Further, social class, gender, race, ethnicity, culture, and religion often interact to make African Americans and Native Americans less likely to trust physicians and other health care providers.

At the micro level, we can use symbolic interactionism to understand why gender, age, race, ethnicity and social class influence health care access and utilization. There is nothing innate in women, nonwhites or people of the lower social class that makes them less worthy of jobs or of health care or any other social resource. Instead, these outcomes occur due to our social construction of reality. For example, being a woman or a man means something to most of us, and we give life to these meanings through our everyday actions. Maybe you do not speak up with a male physician because you identify yourself as feminine, and therefore value men and their actions more than women. Or. maybe, as a young person, you give in to social pressures to conform and experiment with drugs or unprotected sex that may result in illness, even death. From these examples we see that interactionist theory breaks down large macro and meso patterns into our behavior in everyday life. Using symbolic interactionism, we see that social institutions, such as medicine, reflect aggregated patterns of our values, attitudes, and behaviors in our everyday life. Accordingly, these institutions will only change if we consciously change the way we interact with others and ourselves day in and day out.

The concept of social construction can also help us understand changes in our health behaviors. What behaviors do we now consider unhealthy that we used to consider healthy? Some answers include smoking and unprotected sex. How about behaviors that we now consider healthy that we used to consider unhealthy or unnecessary? We could respond with, among others, masturbation, drinking red wine, eating "good" fats, and wearing seat belts and bike helmets. So, thinking sociologically, we should also take notice that the health behaviors of any particular time or place represent the common beliefs and norms of that time and place.

TASKS

1. You work for a senator who wants to support a bill to begin developing a universal, national health care program. As her research assistant, she needs you to write a report on why the U.S. does not socialize medicine and how such a program would positively and negatively influence the U.S. culture and social institutions. As a busy professional, she needs a short report, called a brief, of only two to three pages.

2. A regional insurance company receives complaints from several major employers about the insurance company's failure to cover medical services, labs, or medications to treat alcoholism. Many of the employers experience high employee absenteeism due to problems with alcohol. Employee absenteeism increases employer production costs, which makes them less competitive in a free, capitalist market. The employers argue that the medical community defines alcoholism as a disease, and thus insurance companies should cover it under their **comprehensive health care** plans. The employers threaten to drop the insurance company from their list of health insurance plans offered to their employees. (Insurance companies sell packages to employers, who then pass on the cost to employees as a "benefit" via premiums, lower raises, etc.) As a mediator hired by the employers and the insurance company, prepare for your first joint meeting with each party by identifying explanations in support of seeing alcoholism as a psychological or biological disease, and explanations that support seeing alcoholism as a social problem. According to both explanations, why does alcoholism occur, and how should employers and insurance companies respond to it?

3. As a citizen preparing to vote in the next presidential election, you wish to learn more about the rising cost of health care, a central issue in this year's election. Each candidate supports one or more of the ideas below to reduce health care costs. Sociologically, which idea or ideas seem the most sound? Which groups in our society would each idea hurt? Which groups in our society would each idea help? How might we implement and manage programs generated from each idea? What kind of social change would we need to make programs generated from each idea successful?

- Revoke Medicare or seriously limit its coverage.
- Detach insurance from jobs, and instead organize and distribute health insurance like car insurance.
- Create more **urgent care** centers and expand the office hours of **primary care** practices to evening and weekends to detour people from using expensive emergency departments for cases that do not require **emergent care**.
- Decriminalize euthanasia, based on data that shows we, as a country, spend large amounts of medical resources to provide life support to patients who do not wish to receive such resources.
- Rationing health care, based on data that shows we spend large amounts of medical resources on patients who die within a short time of receiving health care resources.

Chapter 12. Power, Politics and Authority

QUICK START

In this chapter, you will learn:

- To identify the informal and formal nature of power.
- How to locate power, authority, and influence in a social setting.
- The elements of leadership.

TERMS

Authority	The legitimate use of power.
Corporate Culture	The way of life that exists within corporation or organization.
Influence	The ability to sway or pressure the outcome of a human interaction.
Leadership	Effectively guiding social interaction to achieve a goal.
Networks	Connections among social actors and/or social units.
Power	The social resource of having control over other people's attitudes, beliefs, values, behaviors, and experiences.
Prestige	The social resource of having social status, such as that of a priest.

TOPICS

Who really has control over your life? You would probably like to say that "you" and "only you" have control over your life. If you think about this for a minute, many forces **influence** or outright control what we do and how we do it. Who or what has **power** and how much do you have? Power is an illusive thing. It only reveals itself when it is used. Let's investigate this.

Sometimes we consciously and systemically give other people power or **leadership** over us. When we do this we authorize persons or persons in a given role and status the right to make decisions for us. Because this process is socially agreed upon, we contend that these persons have **authority**. We collectively or personally agree that this legitimate use of power is acceptable. Other times, we find ourselves in situations in which forces, that we neither authorized nor feel we have control over, invoke power on us.

A formal way that we assign leadership and authority is through **government**. There are several types of government including **democracy**, **monarchy**, **authoritarianism**, and **totalitarianism**. In a democracy, all of the people have power, usually via voting for elected officials. However not everyone equally participates in voting for officials or in running for office. For example, in the U.S. more white and older people vote than other groups. For the most part, only people with large sums of money at their disposal run for elected positions.

In a monarchy, one family has sole power and this power is passed down to each generation of the family. This family makes all decisions and the people have little say in the government. An example of a monarchy is the British monarchy of the 19th century. Today the Windsor family no longer runs Great Britain exclusively. In authoritarianism, the people have no say in the government, and once in office, there is no diplomatic way to remove leaders. Leaders tend to gain power via force. Totalitarianism goes a step further and dictates what people can and cannot do, and silences any voiced opposition to the government. Examples of authoritarianism include Singapore, Kuwait, and Ethiopia. Examples of totalitarianism include Nazi Germany, China, North Korea, and the former Soviet Union.

For the remainder of this chapter we'll focus on power and "politics" as they appear in daily life and organizations. An understanding of national and international politics is clearly the focus of political science, and we encourage you to investigate this important discipline. We'll be looking at "politics" with a small "p," such as "corporate politics" or "community politics."

Power is formal and informal. It is created and maintained in social systems. Sometimes government and laws, corporate policy, or regulations dictate the formal nature of who controls whom. Power is also emergent. That is, it evolves from the interactions of persons in a group, and it exists within and sometimes in spite of the formal power structure of that group.

Who has power? It's probably better to ask, "what" has power? We often impute power to individuals. However, groups and organizations, and positions within groups and organizations, have power in and of themselves. In other words, the social order, not an individual, conveys direction and control. **Corporate cultures,** ways of life that exist within a corporation or organization, exemplify this. Positions within some corporate cultures have power, but powerful positions also have **prestige**. Prestige provides a social resource of having social status.

Let's think about this for a minute. Remember that the roles that you play and their status are sets of expectations and obligations for your activity. Hence, they dictate what "you're supposed to do." This suggests restrictions on your freedom, and for the most part, your use of power. While your supervisor's personality may have something to do with how this power is used, the main reason that she can tell you what to do -- and, the reason you do it (most of the time) -- is structural. In a formally organized vertical power structure the social system is designed to allocate resources in line with one's position in the "up and down" dimension.

Vertical Power Structure

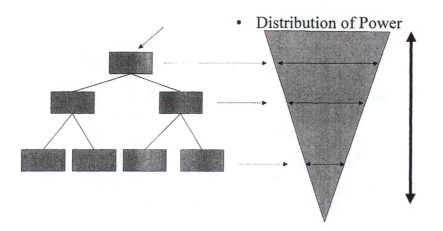

- Distribution of Power

By resources we mean, for example, access to information, control over employees, means of producing things, and the ability to make decisions. This has little to do with whether or not you're a "nice person." Important here is the notion that this is a "created social reality." It can be restructured and recreated. By definition your behavior is, at the very least, controlled by your position. How does this happen? If we borrow a functionalist view of this allocation of power, "it happens" because the social system's configuration generally serves some societal need. Needless to say, you can see how conflict views of this same system suggest that exploitation may well be "structured into" the system. We'll return to this shortly.

Vertical distribution of power begs the question: "So, is there a horizontal power structure?" Sure. But consider some possible outcomes. Reasonably, roles at about the "same" level in the

Horizontal Power Structure

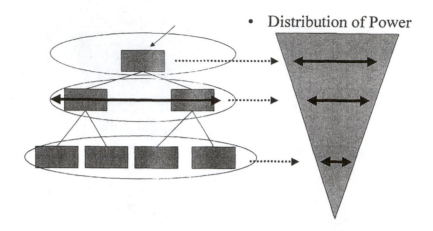

- Distribution of Power

organization have similar levels of responsibility, power, and authority. In a functionalist world efficiency is gained from cooperation among roles at common levels to "get a job done" which in general serves the company's mission. But without carefully maintaining these relationships, the organizational structure produces a situation in which roles of relatively equal power and authority levels compete, perhaps conflict, with one another over scarce resources. Many of us have been involved in what some people call "turf wars"! These are battles over the definition of what responsibility, resources, and power belongs to whom.

Let's take this one step further. If by now you don't believe that social structures intrinsically allocate power, consider organizational "downsizing." Many organizations do this to increase

Organization Before Downsizing   Organization After Downsizing

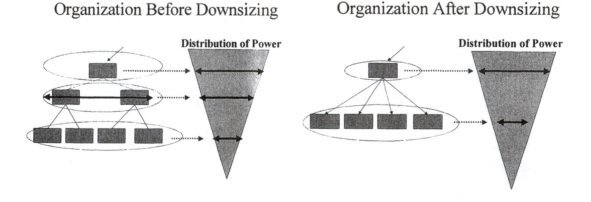

efficiency and productivity. Clearly, persons often lose their jobs and the remaining jobs are redefined. With fewer people doing more work, productivity increases. One notes that not only are the roles and statuses redefined, but also power and authority are redistributed. We often hear of "empowering workers" after a restructuring. You can see how this happens as a matter of organizational design.

From this example, you become aware that sometimes you can't directly control the outcomes of a situation. However, that does not mean you are completely without the ability to sway or pressure others to consider a change of direction. In this case you have no direct power, but you do have influence, and the potential to have impact through others. We've all been told, "It's not what you know, but whom you know that counts." This is not totally accurate -- what you know is very important in a knowledge-based society. But your connections to others through **networks** are important. These networks are often formal, but some of the most influential may be informal. Your social connections to people with power or prestige may give you indirect access to enhanced opportunity, resources, or information. The Internet is a simple but power example of this. Your ability to electronically network to information that before was only in the hands of professionals enhances your power as an individual social actor.

TOOLS

Often it is valuable to assess a social situation by addressing the forces that influence it. Applying sociology often means addressing the nature and relative strength of social forces. If we can estimate "what, who, and how much" each force brings to bear on the situation, we may be able to assess the direction a social relationship will go. A classic technique to visualize this is the "force field analysis" created by social psychologist Kurt Lewin. This tool helps determine the nature and value of the forces and their potential interactions.

Force Field Analysis[9]
A force field analysis is useful in determining social groups that oppose one another, particularly when identifying a target solution. While it is clearly useful in applying conflict theory it can be valuable from a functionalist view -- showing the tension between system forces and needs.
You can do this by yourself or in-groups. Groups open up possibilities for a broader range of thoughts.   1. Draw a large letter "T" on a flip chart or a piece of paper.   2. At the top of the "T," write the problem that you plan to investigate.   3. On your right, write the ideal situation that you're working toward.   4. Through brainstorming, on the left side of the "T" identify the external and internal forces that are driving you toward the ideal situation.   5. Through brainstorming, on the right side of the "T" identify the internal and external forces that are operating against achieving the ideal situation.   6. Prioritize the items from above that drive you toward and away from the ideal situation.

---

[9] Adapted from Brassard, M. and Ritter, D. (Eds.) 1994. *The Memory Jogger II*. GOAL/QPC: Methuen, Massachusetts. Pages 63 and 64.

Another tool that sociologists use is measuring political preference, which ranges from conservative to liberal on various political issues such as the death penalty, abortion, and military spending. The General Social Survey[10] measures political preference as follows:

"We hear a lot of talk these days about liberals and conservatives. I'm going to show you a seven-point scale on which the political views that people hold are arranged from extremely liberal – point 1 – to extremely conservative – point 7. Where would you place yourself on this scale?"

Category and Code	Percent Response in 2000 Survey
1. Extremely liberal	3.80
2. Liberal	10.93
3. Slightly liberal	10.12
4. Middle of the road	37.42
5. Slightly conservative	13.84
6. Conservative	14.59
7. Extremely conservative	3.16
8. Don't know	5.29
9. No answer	0.85

TASKS

1. Select an organization or group in which you are member (work, school, church, or family, for example). Now locate (if it exists) the formal organizational chart. From this chart, write down the formal leadership positions. Now identify 3-4 people who "really are in charge." Are the formal and informal power structures identical? How are they different? Why?

2. Select a change that you want to make in a group or organization. Identify the person or role that must be influenced to make this change. Then trace the connections (network) from you to this person or role.

3. You have been asked to lead a community group. Outline the strategy that you will use to take on this leadership role.

4. Map your spheres of influence. That is, in what ways are you able to sway or pressure the outcome of a human interaction? How can you extend this influence?

---

[10] From James Allan Davis and Tom W. Smith. General Social Surveys, 1972-2000. Principal Investigator, James A. Davis; Director and Co-Principal Investigator, Tom W. Smith; Co-Principal Investigator, Peter V. Marsden, NORC ed. Chicago: National Opinion Research Center, producer, 2000; Storrs, CT: The Roper Center for Public Opinion Research, University of Connecticut, distributor. Page 96.

Chapter 13. Population and Human Ecology

QUICK START

In this chapter, you will learn:

- The characteristics of human population.
- Some frequently used ways to categorize demographic information.
- The use of POET as a powerful analysis tool.
- The value of applying demographic thinking to social settings, that is, social demography.
- Ways to calculate frequently used demographic statistics.

TERMS

Age Distribution	All the different ages represented in a population, listed in numerical order, along with how many times each age occurs.
Age Specific Birth Rate	The amount of babies likely to be born to a particular age cohort.
Age Specific Death Rate	The amount of deaths rate likely occur among a particular age cohort.
Census	A counting of a population's size and characteristics, obtained through surveying or interviewing each adult member of the population.
Cohort	A group of people who share a similar birth year, or share a similar historical experience.
Crude Birth Rate	The amount of babies likely to be born in an area or population, not taking into account the number women of reproductive ages "at risk" for giving birth.
Crude Death Rate	The amount of deaths likely to occur in an area or population, not taking into account the sex or the age of the area's residents.
Crude Growth Rate	The amount that an area's population size is expected to increase
Demographic	Individual, group, or societal characteristics that reflect core distinguishing characteristics of human populations. Age and sex are commonly used examples.
Emigration	Population movement out of a country or area.
Fertility	The ability to give birth, and the characteristics associated with it.
Global Economy	Cross culturally and internationally interdependent economic system that ties markets, production, labor forces, and distribution into a worldwide system.
Human Ecology	The interaction of collective human action and the environment.
Immigration	Population movement into a country or area.
Median Age	The point of a group's age distribution at which 50% of the population is younger and 50% of the population is older.
Population Size	The number of people who live in an area.
Sex Distribution	The number of men and women in a population.
Sex Ratio	The number of men per 100 women.
Migration	Population movement into and out of a country or area.
Mortality	Death and the characteristics associated with it.
Population	The number and "kind" of people that are determined to exist in a socially defined location at a given point in time.
Sex Specific Birth Rate	The number of babies likely to be born to a group of people.

		The number of deaths likely to occur among a group of people.
Sex Specific Death Rate		The number of deaths likely to occur among a group of people.
Snowbirds		People who temporarily migrate to warmer areas during the winter.
Social Demography		The social analysis of population characteristics such as a group's size, growth rate, age, sex, race, ethnicity, sexuality, income, education, religion, language, marital status, and employment.
Sunbelt		The strip of mostly Southern states with warm climates year round.
Vital Registration Systems		Data documenting all births, deaths, marriages, and divorces in an area.
World Growth Rate		The amount that the world's population size is expected to grow.

TOPICS

There's little doubt that we have **population** issues on earth. Usually, a discussion of them centers on the earth's growing population size, with the **world growth rate** larger than the earth can stand. A growing population often results in serious environmental and public health problems such as poor water quality, waste disposal problems, starvation, and disease. We sometimes ignore population growth problems reported in the media because they seem so far away. But an understanding of general population issues is vital in a **global economy**. In this chapter, we aim to help you develop such an understanding by exploring the impact of social forces on population and the influence of population on society.

One way to view the impact of population change is through a social ecological approach. This approach is extremely versatile and can be used to conceptualize or "get hold of" some complicated social situation. A simple, but powerful way to think about this is to use the acronym: POET (population, organization, environment, and technology). Each of these is a major force in its own right, but your ability to analyze complicated social situations is enhanced by looking at the interaction among these forces and the likely impacts that these interactions produce. The table below summarizes the definitions and interactions among these forces.

	Population	Organization	Environment	Technology
Population (how many people and what kind)				→ ↓ a
Organization (how a social system is structured)		↓		
Environment (physical environment, the material culture, too)	→	b		
Technology (collective human "know-how")				

Let's take two quick examples of how these forces influence one another. Cell "a" above represents the impact of technology on population. In the last half of the 20th century, collective human "know-how" produced birth control pills. The impact on population was a reduction in the number of births. Other forces influenced such an impact. Some religious groups (organization) opposed the invention, while health care, economic, and governmental groups

86

supported the creation and delivery of this technological breakthrough (organizational). Family size was altered and gender roles were influenced (organizational). Declining **fertility** meant potentially less impact on the physical environment (environment).

Let's move to cell "b." This time, let's look at a meso level: an organizational level. Suppose an information technologies company plans to expand its work force. Expansion of work force may mean increased expansion of office space to accommodate new employees. More office space means more environmental impact. How the company decides to organize its work and its technology can have a varied effect on the environment. If the company decided to build new office space at a fixed site, this will mean more commuters, more automobile miles driven, larger parking lots. In short: more direct wear and tear on the environment. Of course, alternative organization of work might preclude the use of new physical space at all. By using computer technology, reorganizing home life and home environment to include work and work tools, the environmental impacts at the very least could be relocated, at best minimized. The opportunity for the use of the sociological imagination is endless!

Now let's focus on population. Sociologists conduct **social demography** to help communities and societies, even organizations plan for the need for social, environmental, and economic resources such as jobs, housing, transportation, schools, hospitals and health clinics, criminal justice, fire departments, waste removal, parks and recreation, and day care centers, to name a few. Population size drives the need for these resources. To effectively plan for the production, distribution, and consumption of key resources, societies must estimate their population size. To estimate population size, we need to now the number of births, deaths, and total **migration**.

Hence, most societies continuously and systematically collect data on births, deaths, **immigration,** and **emigration**. In the U.S., we record all births and deaths and all border entrances and exits. We collect this data through a massive undertaking at the local, state, and national level, including a **census** each decade, **vital registration systems**, and numerous face-to-face, telephone, and mail surveys. From these primary sources, we also collect various **demographic** data on residents such as sex, race, age, education, marital status, children's ages, employment status, and income. We use numerous secondary sources to obtain other population data, such as tax forms, property transactions, W-2 income forms, and the amount of new and existing housing in an area. Sociologists possess unique tools and training to conduct social demography. Subsequently many sociologists at the bachelor's, master's, and doctoral levels work for local, state, and national governments collecting and analyzing population data.

Now that you know what social demography is, let's dig in a littler deeper on the three main components of population estimates -- births, deaths, and migration. When the number of births increases, a population becomes younger, because more young people join the total population. When the number of births decreases, a population becomes older, because fewer young people join the total population. Babies are the most vulnerable to **mortality**. Hence, as the number of deaths increases, a population becomes older, because more babies die. As the number of deaths decreases, a population becomes younger, because more babies live.

As for migration, you probably know that as it increases, the population size increases. As emigration increases, the population size decreases. Historically, the U.S. experiences high immigration and low emigration. Migration affects more than population size, though. It also influences the demographics of an area. Young and senior adults migrate more than other age groups. Young adults leave areas for better jobs or further education. Many are not yet settled with careers, spouses, children, or homes, making it easier for them to migrate. Older people leave areas for retirement opportunities, such as desirable housing, proximity to adult children and their families, mild climates, and health and social resources such as hospitals, medical specialists, and recreational amenities.

Migrants of all ages are likely to be relatively wealthy. Why? Because it takes money to move. You need immediate funds to pay for the move itself, new housing, and to offset a potential lag in pay from a new job. Further, you need leisure time away from work to find new housing. Subsequently, emigration leaves an area older and poorer, with a smaller labor force and a lower tax base to fund schools, roads, police, medical care, and so on. Areas experiencing extreme emigration, like towns across West Virginia, often become ghost towns.

In contrast, immigration often brings money and the need for additional resources such as jobs, housing, shopping, schools, health care, police and fire services, and so on. For these reasons, many areas try to attract migrants by creating better schools, hospitals, public transportation, recreation, and so on. These resources are expensive to areas, though, requiring the investment of higher taxes for years in exchange for the uncertainty of future migrants.

However, people temporarily relocate, such as those who move south in the winter and north in the summer, allow areas to reap the benefits of immigration without the costs. **Snowbirds** provide economic benefits such as increased purchasing of local goods and services and higher sales tax revenues, without the costly investments in new or better social resources such as permanent housing, jobs, schools, expansion of police and fire departments. For these reasons, the many locales in the **Sunbelt**, such as the state of Florida, strategically attract snowbirds. Many areas also try to attract large companies to locate or relocate to their area because they bring jobs, as well as higher revenues from income taxes, property taxes, sales taxes, and corporate taxes, which fund schools, roads, libraries, parks, museums, and so on. If occurring unexpectedly or too fast though, immigration burdens an area. Hence, areas need to predict population changes, and plan accordingly. This is where social demographers come in. They use the tools discussed so far in this book, as well as specific demographic tools. Let's take a look.

TOOLS

We can simplify social demography by applying two sets of tools to population questions. We need to ask and answer "how many?" and "what kind of people?" types of questions. We will start with the "how many" tools. Below is a general formula to calculate population size:

Population Size = previous # residents + # births - # deaths + # immigrants - # emigrants

Let's walk through a quick, simple application of the above formula. Suppose a group of 20 Italians emigrate to the Kansas City area. Over the last year, one Italian dies, five emigrate back to Italy, four new Italians join the group, and two Italians each give birth to a baby. Using the formula above, we can calculate the population size of the group at the end of the first year:

General Population Size = 20 + 2 - 1 + 4 - 5

General Population Size = 20

Now it is five years later, the group includes 25 more Italian immigrants with ten more births and five deaths. With this data, we can calculate a **crude birth rate** and **crude death rate**. These rates are called "crude" because they do not include sex or age data. The "risk" of birth is not evenly distributed across men and women of different ages. Women's most fertile ages are 15-49, with peak fertility from 20-29. Men stay fertile for most, if not all, of their life span. Similarly, the risk of death is not evenly distributed across men and women of different ages. Babies are the most likely to die, followed by elderly people. That having been said, let's apply the formulas below to calculate the Italian community's crude birth rate and crude death rate:

Crude Birth Rate = [(total # of births / total population size) 100] / # of years

Crude Birth Rate = [(10 / 20) * 100] / 5

Crude Birth Rate = 10 per year or 50 over five years

Crude Death Rate = [(total # of deaths / total population size) 100] / # of years

Crude Death Rate = [(5 / 20) * 100] / 5

Crude Death Rate = 5 per year or 25 over five years

Now the Italians want to predict the future size of their community so that they can plan for necessary social resources. To do so, we need to calculate the growth rate and then use it to project future population size. A positive growth rate means the population is increasing, a negative growth rate means the population is decreasing, and a growth rate of 0 means that the population is staying the same size, with an equal number of births and deaths. We use the following formula to calculate the **crude growth rate**:

Crude Growth Rate = {[(current population size - previous population size) / previous population size] * 100} / # of years

Crude Growth Rate = {[(25 – 20)/20] * 100}/5
Crude Growth Rate = 5 per year

The community's current growth rate equals 5 a year, for a total of 25 over the five years. Using the growth rate, we can estimate the population size in 5, 10, or 15 more years from now:

5 Year Projected Population Size = [(current population size * growth rate) /100] # of years of projection + current population size

5 Year Projected Population Size = [(25 * 5) /100] 5 + 25

5 Year Projected Population Size = 31.25

Knowing future population size and demographics enables groups to conduct community planning. Because growth is occurring mostly from births, the community leaders can expect to need more resources for children such as day care facilities and a playground. In general, a failure to plan can create a scarcity of necessary social resources, which may result in a multitude of social problems. This is happening today in areas experiencing quick and unexpected growth from retirees or Latino immigrants.

Next, we need to ask the "what kind of people?" types of questions. Here we are interested in the demographic characteristics of populations, not just the size of populations. Knowing the **age** or **sex distribution** of a population enables us to interpret the validity of our crude growth rate. In the Italian community example above, we estimated a crude growth rate of 5 a year. What if we knew the age and sex distribution of the group members in addition to the one year and five year population size? Let's say the ages of the adults in the community range from 18 and 45, with an average age of 24.5. The adult sex distribution is about 50 men and 50 women.

What does knowing these demographic characteristics tell us? Given equal numbers of men and women, all of peak reproductive ages, we can predict that 5 year crude growth rate of 5 would likely remain constant for the next 15 years. At that time, the neighborhood would begin aging out of peak reproductive years, reducing the growth rate. Now, what if nearly all of the residents are men, say 90? In this case, we would expect the growth rate to decrease immediately and eventually become negative.

Knowing age or sex distribution of a population also helps us refine our crude birth and death rates into **age** and **sex specific birth** and **death rates**. These rates provide more valid estimates because they take into account the varying risk for birth, death, and growth across men and women of different ages. Below are some example formulas for age and sex specific birth and death rates. You can see from these formulas that the resulting birth and death rates will be much more specific. You can practice plugging in different hypothetical numbers and see how the age and sex specific birth and death rates differ from the crude birth and death rates.

Birth Rate for Women Age 15-19 = [(total # of births among women age 15-19/ total # of women age 15-19) *100] / # of years

Birth Rate for Women Age 20-24 = [(total # of births among women age 20-24 / total # of women age 20-24) *100] / # of years

Death Rate for People Age 0-4 = [(total # of deaths among people age 0-4/ total # of people age 0-4) *100] / # of years

Death Rate for People Age 75-79 = [(total # deaths among people age 75-79/ total # of people age 75-79) *100] / # of years

Now let's look at two more uses of age and sex data in social demography. Sociologists can use a population's **median age** to estimate the population's growth. When the median age is approximately 35, countries achieve zero population growth. Countries with low median ages, such as developing countries, have high population growth rates. In contrast, countries with high median ages have negative growth rates.

Sociologists can use a **sex ratio** to predict social change and varying population needs. A sex ratio is the number of men per 100 women. A sex ratio of 100 means that equal numbers of men and women exist. How do sex ratios affect social structure? A sex ratio of over 100 means a population contains more men than women. This often occurs in military towns and areas with large numbers of physical labor jobs, such as towns across Alaska now, and towns across the American west at the turn of the last century. With a high sex ratio, we expect towns to contain a higher number of bars, strip clubs, massage parlors, prostitution, fast food restaurants, transient residents, and a higher demand for inexpensive, temporary housing. We might also expect a lower number of churches and fewer schools and day care centers.

A sex ratio of less than 100 means a population contains more women than men. This happens frequently for many reasons. To begin, more females survive the first year of birth, leaving a lower than 100 sex ratio in the early years of a **cohort**. Due to wars, the sex ratio often becomes even lower as a cohort reaches ages 18-25. Historically, the sex ratio drops even further across the life span, because, on average, men die at younger ages then women. This occurs because more men than women work in risky occupations, and historically more men drink and smoke.

What happens when a population consistently contains more women than men in each cohort? Many cultural and structural changes could occur, such as fewer marriages and more single adult women. More marriages might occur between younger men and older women. With time, we might no longer expect or plan to marry or have children, dramatically changing our socialization processes and outcomes. The stigma of divorce could dissolve. The rules of social interaction could also change, with people becoming more sexually permissive as they no longer wait for the "benefit" of marriage. More babies might be born outside of marriage. Definitions of female beauty might move away from the tall, thin, and young, and instead include women of all sizes and ages. Further, with a lower sex ratio, more women might seek full-time employment and professional careers, becoming economically independent.

TASKS

1. A northern city is experiencing an increasing rate of Latino immigration. City leaders want to identify what changes they can expect in community needs and characteristics. You work as a community planner for the city. Using your sociological training, identify how an increasing Latino population may change the city's demographic characteristics and change the needs of each social institution including housing, employment, crime and deviance, health, marriage and family, education, and religion.

2. The department of travel and tourism in the state of Arizona wants to attract more retirees to visit or move to Arizona. It announces a request for proposals (RFPs) from social science researchers and consultants on plans to achieve this goal. You own a consulting firm and plan to submit a proposal. What will be the main components of your proposal?

3. You work as a social demographer for the museum of American history in Washington, DC. You need to create a storyboard on how the important demographic cohorts across the 20th century shaped American culture and society. Use the terms, topics, and tools in this chapter to draft a storyboard. Be sure to include statistical data such as sex ratios, birth rates, death rates, and possibly charts to demonstrate the trends. Below is a list of the cohorts:

   - Good Times (born in 1930s)
   - Baby Boom
   - Generation X
   - Baby Bust (Baby Boomers' children, born 1976 to today)

Chapter 14. Social Change and the Future

QUICK START

In this chapter, you will learn:

- To recognize the extent to which your worldview is culture bound.
- To recognize and use paradigms to understand present and future social reality.
- To recognize why it is important to think about the future.
- To effectively create a vision of the future to guide human action.

TERMS

Culture Bound	Being tied to the culturally learned definition of reality, space, and time that both grounds and limits our ability to view reality.
Driving Forces	A set of social forces with the greatest impact on a social problem.
Future	A time and state of being beyond the present.
Paradigm	A collective way of thinking and knowing the world, which frames and limits the way we see reality.
Scenario	A proposed view of the ways things might be in the future.
Think Inside the Box	Ideas about a problem that are determined by the social situation in which the problem exists, and by how others have thought in the past about the problem.
Think Outside the Box	Ideas about a problem that are not bound to place, time, or people.

TOPICS

Are you different today than you were yesterday? Whether you recognize it or not, things probably are different than they were. The social context in which you live has changed and you have changed. How is the social context different today than yesterday, last week, or last year? To what degree are you different as a result of these contextual changes? Earlier we addressed the value of using and creating social theories as tools. These theories help us explain this change. Here we'd like to view social change by looking at the future.

Why think about the **future** anyway? First, on a philosophical level, as social selves, we are comprised of our past, present, and future. In other words, in order to interact today at work, with our families, and in our communities, we often need to know what happened yesterday, and anticipate what will happen tomorrow. Second, we can't get there -- the future -- if we don't know where we want to go. How do we want our future society to look? How do we want it to operate? What will the social roles, norms, and institutions be? To guide us, we will need to create a vision for our future society. Never fear! Our sociological tools will help us to better construct our vision for the future. Third, a vision of the future acts as a magnet for our action. A vision of the future helps us determine where to put our time, our energy, and our resources.

In this chapter, we will look at thinking in the present and the future, which basically translates into thinking "inside" and "outside the box." You have probably heard people say that to be successful today and in the future we need to **think outside the box**! If you're like most people, you probably wonder, "What is the box? Where is the box? What does it mean to think outside the box?" Sociology can help you answer these questions.

Have you ever literally been inside a box? Maybe when you were a child you crawled inside a box. What was it like? Regardless of its size or shape, a box confined you, limiting your movement. This confinement might have been comfortable for you, creating a sense of safety and security, especially if you were in trouble with a parent or older sibling! In another setting, being confined in a box might have made you scared or unsettled. What if you believed that you could not get out of the box? In that case, feeling scared and unsettled might progress to panic. Thanks to the self-fulfilling prophecy, your breathing may become heavy and your pulse fast.

Now that we have explored what it is like to be physically in a box, let's look at being symbolically in a box. What does it mean to **think inside the box**? It means your thinking is determined by the social situation in which the box exists, or by what you personally bring to a situation. With regard to the latter, our psychology friends would no doubt agree that individuals create their own boxes from their fear, personality, or perception. More sociologically, our internalized norms, values, beliefs, and ways of seeing the world comprise boxes.

Sociologists refer to widespread sets of internalized beliefs as **paradigms** (and we don't mean a "nickel short of a quarter"). Paradigms direct how groups of people think and act. As such, a paradigm is a box. For example, in the past, people believed the world was flat. People thought it risky to travel far in any direction on the sea for fear of falling off the edge of the world. This belief was a box, limiting how societies grew and interacted, shaping social institutions such as education, marriage and family, religion, science, politics, the military, and the economy. We can think of the belief that the world was flat as a paradigm because it influenced nearly every aspect of social life.

How about something more modern? Getting information and transmitting it has become commonplace. To put it to use, our minds still need to process the information and make sense of it. Suppose we want to skip this step. We want to download information directly into our minds. This is the logic behind virtual reality, and the premise to the hit movie *The Matrix*. Most of you are probably thinking "No way! It can't happen." If so, you are thinking in a box!

Computers create electrical impulses. Similarly, our minds create electrical impulses called thoughts. So, if we can transfer electrical impulses from one computer to another, why can't we transfer them from a computer to our mind? Think of how this technology would change our lives. We could travel anywhere, and learn anything. If we create this technology, we probably would eventually be able to upload information from our minds to a computer. Actually, the technology to teach computers to think recently became available. Obviously, we need to get a handle on the ethical and social implications of this technology. Teaching computers to think may have many undesirable outcomes, as portrayed in the movie *The Matrix*.

In sum, these propositions may be overwhelming to us now, but conceptually and technologically, they are possible. If you think these changes will never happen because we've never done it that way, you are thinking inside the box. Thinking about the future almost certainly requires that we think "outside the box."

Let's expand this idea of a paradigm as a box. Sociologists often ask questions like, "What will the American family look like in 2050?" Don't answer, it's a trap! It's a cultural box. You may think that the fact that we're looking a half-century down the road gets us "outside the box." After all, nobody knows for sure what the family will look like in the future. Besides, that's a long way away! Here's the catch. When we pose this question we reflect on our cultural boxes. We make assumptions about the future based on our current and past definition of reality, which we've learned and internalized through social interaction. We bring our learned "baggage" to just about any social problem or question. For example, the notion of family itself is a box. We learned the very idea and meaning of the word *family* through socialization. To varying degrees we live by our collective idea of family. Sure your perception and experience of family may have some unique elements to it. But you also live in a society in which shared beliefs give rise to similar behaviors and expectations, creating more shared perceptions and experiences than unique ones. Furthermore, we learn the meaning of our individual perceptions and experiences through social interaction. In turn, we use these definitions to guide social interaction. Thus, the social construction of reality shapes the way we live and think, often becoming a box.

Don't misunderstand; this cultural box isn't necessarily a "bad" thing. Boxes provide structure to daily life, making life easier to live. Living outside the box can be a very unpleasant experience. Imagine needing to consciously define "taken for granted" perceptions, behaviors, and expectations every moment! As an example, just think how hard it would be to drive on a busy road in a town or city without the "boxes" of streetlights, signs, and white lines! So as you swim the channel of life, a box may be a conceptual weight around your leg, and a heavy one at that, or a structural life raft that keeps you afloat.

TOOLS

How can we use this notion of thinking in and outside the box as a tool? The theory of symbolic interaction helps us here. According to symbolic interaction, we are **culture bound** by time, space, and symbols. Hence, we experience real trouble thinking the unthinkable. Here's an example. Try to answer this question: "When will the family cease to exist?" We've asked this question hundreds of times in our years of teaching. While we haven't collected any quantitative data, our on-board personal computers (i.e. our brains) tell us that the most frequent answer is "Never." Looks of bewilderment, anxiety, and sometimes anger usually follow this answer in class. A number of people are simply uncomfortable by the question. Families are important, and we believe they have been around for as long as humans have inhabited the planet. Many of us even think that the family is sacred. How can we imagine that it won't exist? To get outside the box, though, we must "think the unthinkable."

Take a look at the boxes below regarding the future of the family as a social institution. This is an oversimplification, but it illustrates how to think outside the box.

Box A	Box B
The family will exist in the future.	The family will not exist in the future.

Do we agree with Box A or Box B? All of us would probably support the notion that the family exists now. We have empirical, experiential, and cultural support for this assertion. If we think linearly, that is, that tomorrow will be pretty much like today, then we would expect the social institution of the family to exist in the future. We would conclude with Box A. In doing so, we remain in the box, within the safety of "the known."

So how do we get outside the box? In reality there are many ways to do this, but in this example there is only one. It is to go into the unknown and assert that the family will not exist in the future. Force yourself to imagine a situation in which the family, for only argument's sake, does not exist. You might say, "Well concluding that doesn't get us outside the box! We remain in a box, maybe a different box, but a box nonetheless." That is exactly the point we are trying to make with the diagram above. The truth of the matter here is that we are never free of the many social influences in our lives or our minds. So you will likely never get completely outside the box, outside history, outside today, outside internalized beliefs, attitudes, values, and norms.

In most cases, thinking outside the box is more of a struggle with the box than it is a transcending or abandoning of the box. It is not a total disconnect from what you know and believe. We need these to imagine what other forces might be in play in the future. Instead of staying moored to a conceptual anchor, we cut the line and disassemble the boat. What questions do we need answers to now? Start by identifying what answers the family provides now. Families provide nurture, protection, nutrition, education, recreation, socialization, and social status. How will these get done in the future? Who or what will do these things?

Now that we've asked the questions, we can begin to use our newfound ability to construct a new social reality. Here you begin to see the value of a little creativity and imagination in what we do. Of course, C. Wright Mills's sociological imagination will need to be expanded in the 21st century. Perhaps we need to move a step beyond imagination and develop sociological creativity. This allows us to move from *thinking* about things as *they might be* to the use of the sociological perspective to *create* new things that *never were*. In this way the theories, concepts, and methods we've learned in sociology serve as construction tools and not anchors.

One way to use this perspective in applying sociology is **scenario** building. Check it out.

Tool	What is it? How do I do it?
Scenario building[11]	A **scenario** is a proposed view of the ways things might be in the future. Rather than a wild guess of what things might be like, a scenario systematically identifies social forces that likely may shape the future. In this sense, it constructs a social reality so that decisions can be made accordingly.
What is the question?	First you must decide what question you want to address in the future. For example, "What will family life be like in 20 years?" How about, "What will higher education look like in 10 years?"
What are the key factors?	Next we need to rely on our brainstorming abilities coupled with an information search. Spend some time thinking about the "main things" that may influence your problem in the future. These are "key factors."
What are the larger driving forces?	Use brainstorming to create a list of key factors. Then cut the list down to those that are likely to have the greatest impact. These are the **driving forces** that direct the process. Now you must make some judgments. Two things determine the importance of these driving forces in your scenario: how likely each is and how important is each.
Create a table like this.	Write a driving force in each cell under the column labeled "Driving Forces." {table below}
Evaluate the driving forces.	1. In the column "Importance," rate each driving force on a scale from 0-10, with 0 meaning not important at all and 10 meaning extremely important. 2. Then, in the column "Likelihood," rank the driving forces with the most likely force receiving the highest number, and the second most likely receiving the next highest number, and so on. Use as many numbers as you have forces. That is, if you have 10 forces your ranking can be from 1 –10.
Summarize your findings.	Choose the top 2 driving forces that have the highest importance and the highest likelihood. Write a paragraph about each of these effects with regard to the original question.

Driving Forces	Importance	Likelihood
Driving force 1		
Driving force 2		
Driving force 3		
Driving force 4		
Driving force …		

---

[11] Adapted from the Four Visions Scenario at http://www.si.umich.edu/V2010/scenproc.html and Schwartz, P. 1996. *The Art of the Long View: Planning for the Future in an Uncertain World.* Doubleday: New York.

TASKS

A national women's rights organization wants to prepare men and women for the changes that the 21st century will bring in the roles of gender, family, and work in our lives. They hire you, an applied sociologist, to help them (1) understand the likely changes and to (2) create a pamphlet presenting this information, which the organization will mass disseminate to the general public.

1. Use what you know about sociology to identify what roles gender, family, and work play in our lives today. How will these get done in the future? Who or what will do these things?

2. Use your sociological theory and concepts to critique the roles of gender, family, and work you identified above. What problems exist in gender, family, and work today? Why do these problems exist and how did they come about? Which of these problems will be addressed in the 21st century? How? By whom or by what?

3. Use what you know about technology and social change to identify new aspects of the roles of gender, family, and work in the 21st century. How, whom, and what will these get done?

4. Summarize your answers above into approximately 10 bullet statements. How could you package and word these statements to induce change in readers' attitudes, values, and behavior? It may help here to create a mock pamphlet by folding a sheet of paper into three folds horizontally. Where and how should you lay the bullet statements on the pamphlet? What would be most eye-catching? How should you reword the bullet statements to maximize understanding in a diverse population? What should you put on the front and back fold of the pamphlet? What graphics would attract attention and deliver the desired message? What ethical issues should you consider here? Why? How?

QUICK START

In this chapter, you will learn:

- How to identify skills we learn in sociology and where to use them.
- How to identify job profiles.
- How to find and apply for jobs.
- How to prepare for job interviews.

TERMS

Career Path	The steps taken to reach a particular job in a particular occupation.
Cover Letter	A summary of a person's professional experience and skills written in paragraph form, usually one to two pages long.
Electronic Resume	A resume accessible via the Internet or email.
Informational Interview	A discussion with a person currently holding a job that interests you, in an occupation that interests you.
Internship	A paid or unpaid opportunity to learn about a professional job while performing it, similar to an apprenticeship.
Job Description	The role expectations of a job.
Job Interview	A meeting between an employer, or his or her representatives, and a job candidate to discuss a job opening and the fit between the candidate, the position, and the employer.
Job Market	The job openings, employers, and applicants for a particular occupation at a particular time.
Job Profile	The characteristics of a job, including salary, hours, expectations, skills required, autonomy, prestige, and power.
Master Resume	A resume that lists all of the sections of a resume an all possible content of each section.
Mentor	A professional role model that guides you in your career.
Mentoring	Guiding the career of a person with less professional experience than you.
Networking	Building, or drawing on, the group of family members, coworkers, clergy, neighbors, professors, friends, and associates that you know on a professional or personal basis.
Protégé	The person receiving guidance from a mentor.
Resume	A summary of a person's professional experience and skills written in tabular form, usually one to two pages long.
Service Learning	Short, temporary community activities associated with learning the concepts or theories of an academic course, such as visiting a nursing home for one afternoon.
Volunteering	Helping an organization reach its goals, without receiving monetary payment or holding an official position within the organization.

TOPICS

Talk to your parents and grandparents. They are likely to tell you that they planned to do one or two different work roles in their lifetimes. For example, your grandfather, at different times in his life, may have been a plumber, a soldier, and a manager. Similarly, you will likely perform different work roles across your life, and may change occupations. Because of rapid change in our economy and technology, you will likely change work roles and occupations more times than your parents and grandparents. Bottom line: Be prepared to change jobs!

The reason for this explosion of new and transitioning jobs is simple but powerful: the velocity of social change is fast and getting faster. Jobs are roles. You know that as social institutions like family, government, religion, education, and economics change, the roles, which are connected to them, change. Meanwhile social forces, like technology change and widespread organizational change, cause the invention of new roles and innovation, or transitions of existing roles, and the disappearance of some roles altogether. All this happens at an incredible rate.

So how are you supposed to keep up with these changes? From a functionalist point of view you'll need to resocialize and prepare to change jobs. This ongoing resocialization necessitates life-long learning. Get used to it: You'll be learning for the rest your life. For a while, at least, you are likely to always be "on" the **job market**. In order to strategically navigate the job market, we need to "match" the role expectations of a job (also called a **job description**) to your skills, abilities, knowledge, and personal desires.

Job Profiles

We live in a post-industrial society. Our society emphasizes the manufacturing and marketing of information, or, more generally, symbols. Hence, in this moment in history, most **job profiles** include some element of information processing. All of the sciences prepare students to create, interpret, and evaluate information. What unique skills do you bring to the job market? What jobs do you qualify for? What distinguishes you from other job applicants? A sociological perspective, gained from even just one or two sociology courses, can help here. Most people run into trouble looking critically at themselves and others. Fortunately, this is exactly what sociology trains us to do.

Let's face it, as useful as many of the skills and perspectives that we learn in sociology are, we rarely see job advertisements reading, "Wanted: Sociologist" or even "Wanted: Persons with Sociological Training." This doesn't mean that you can't connect the skills that you've learned in your sociology classes to valuable job skills. We may simply need to translate some of our skills into the areas in which we want find work. Even some sociological training -- one course -- can be valuable to your work experience. Here's an example. Let's take a recent 22-year-old graduate with a bachelor of arts in sociology. What unique skills does he possess? What kind of jobs should he apply for?

Skills We Learn in Sociology	How and Where We Can Use It	
The Sociological Perspective	Counseling	Law Enforcement
	Diversity Training	Management
	Education and Teaching	Marketing
	Government (all levels)	Organizational
	Health Care	Development
	Human Resources	Planning
	Information Technology	Real estate
	Insurance	Retail
	Law	
Research Methods	Advertising	Insurance
	Continuous Improvement	Law Enforcement
	Education and Teaching	Market Research
	Finance	Organizational
	Government (all levels)	Development
	Health Care Research	Needs Assessment
	Human Resources	Program Evaluation
	Information Technology	Real Estate
Sociological Theory	Conflict Resolution	Planning
	Human Resources	Strategic Planning
	Needs Assessment	Systems Analysis
	Organizational Development	

This is just the start. We believe you can see that searching for jobs that have key words from the column on the right in the above table expands the options and increases the possibilities of finding a job related to your skills. Of course, other sociological skills like statistics and problem solving are valuable for almost any job or occupation.

In general, persons with a bachelor's degree in sociology are eligible for most entry level jobs in a wide variety of fields -- business (don't forget retail and finance), government (don't forget local government -- county, city), non-profit organizations and health care (don't forget HMOs and local hospitals) and, of course, education. Research organizations, criminal justice, and journalism are also some initial possibilities (don't forget local police and local newspapers). In searching for a good fit between you and meaningful work, you'll need to use a little self-motivation and ingenuity. Be positive, proactive, and draw on your sociological skills. In the next section, we provide specific tools for finding jobs, writing resumes and cover letters, job interviewing, and negotiating job offers.

TOOLS

Finding Jobs

When considering a job or occupation, you can gather information about the importance of it to society, its demand and supply, as well as its salary, power, and prestige. This is a place where those of us with sociology skills should excel! You can gather information about jobs from the library, bookstore, Internet, or professional associations such as the *Society for Applied Sociology*, as well as local and regional professional associations. While job searching, remember that you will rarely see a "Help Wanted: Sociologist" sign or advertisement. We have skills for sure, but often, particularly at the bachelor's level, they're integrated into job titles that don't simply identify "sociologist."

In *Getting a Head Start on Your Career as an Applied Sociologist* (1998), Catherine Mobley, Steve Steele, and Kathy Rowell discuss some useful strategies that can help you learn about careers, develop skills, gain experience, and perhaps find a job! Below we list some of their strategies. Try combinations, or all of them, simultaneously.

1. *Networking*. Why not use your connections to get a job? "I don't know anybody," you say? That's just not true. Chances are you're not looking at the full breadth of your network. Consider family members, friends, as well as professionals in the field of your interest. Contact them, let them know of your job search, and ask them to "keep their eyes open" for upcoming jobs. They may be able to "hook you up" with an opportunity.

2. *Informational Interviewing*. If you learn of someone who holds a job that interests you, gain firsthand career insight about the job and the occupation by conducting an informational interview. Contact the person who holds the job, explain your interest in the job, and ask to schedule an appointment to talk about his or her job. It is important to convey that you are not requesting a "job interview." Rather, you only seek to learn about the job and the occupation. At least two things can happen if they agree to an appointment: You'll learn whether you really want a job or career like that one, and the interviewee might remember you when a job opens up. Bring your resume, but wait for the interviewee to ask for it. The kinds of questions you might ask include, "How did you find this job?" "What previous jobs did you hold?" "What skills, training, and education are necessary to perform this job?" For more informational interview questions, see Mobley et al.'s (1998) *Getting a Head Start on Your Career as an Applied Sociologist*, and the American Sociological Association's (1998) *Embarking Upon a Career with an Undergraduate Degree in Sociology*.

3. *Mentoring*. Find and use **mentors** at each step or stage of your career. Mentors guide you in the direction you want to go. A mentor is someone from whom you can ask professional, and sometimes, personal advice. A mentor provides proactive guidance, in that he or she does not wait for you to request advice. They may initiate the contact. Often a mentor can suggest career training, identify job opportunities, and "tip you off" to the possible pitfalls along the way. A mentor can pass on some valuable informal skills such as how to network and present

yourself. Establish mentors by approaching role models, as well as people with jobs that you might like. The relationship between a mentor and a **protégé** is mutual and dynamic. A protégé helps the mentor as well as receives help from the mentor. For example, the protégé might provide information or insight on a topic they know more about, or the protégé might assist the mentor with a project. Eventually, the protégé may provide mentoring to the mentor. Usually, mentoring relationships end slowly, as the protégé needs less professional guidance.

4. *Interning*. Get a paid or unpaid **internship**. Interns learn the expectations of a job while they successively develop skills and training for the job. Internships also structurally link interns to occupations, professionals, potential mentors, informational interviews, and networks. Plus, as an intern, you get a chance to show what you can do!

5. *Volunteering*. No money, but some valuable experience and networking can unfold as a result of volunteering. As a volunteer, you learn more skills -- sometimes, advanced skills that you can't learn in school or in paid jobs. Also, you're able to practice these skills. Furthermore, a volunteer experience may lead directly to a paid job inside the volunteer organization, or, through expanded networking, to a paid position outside the organization. Be careful though: It is easy to take on volunteer opportunities. Don't overextend yourself. It will only result in disappointment for everyone involved.

All of the above techniques help you to identify job skills, job qualifications, job titles, professionals, and **career paths** in desired occupations. They may connect you with the job you want! We would never suggest that we have a complete list of "do's and don'ts" for job seekers with sociological skills, but here are a few items that may help.

Job Seeking Approach	Do's	Don'ts
College Job Placement Office	Search for all jobs in all fields with related expectations and skill levels. Identify all the related jobs in a field via the strategies above.	Just look for "Sociologist Wanted" or "Social Science Degree" only.
Internet Job Search Engines	Search by job titles relevant to the field of interest. (Again, you can identify relevant job titles by using the above strategies.) Search regularly to learn job titles and job descriptions.	Just type in "sociologist" or "sociology" and expect hits.
Classified Ads	Look for jobs with skills relevant to your repertoire. Look regularly to learn job titles and job descriptions.	Look only in one occupation or category.
Personal and Professional Networking	Tell everyone about your job search, including family, friends, coworkers, clergy, neighbors, and professors. Join professional organizations such as the Society for Applied Sociology, the Sociological Practice Association, and the American Sociological Association.	Underestimate the importance or effectiveness of networking. Most people obtain jobs through networking.

Regarding Internet searching, you might want to start at our coauthor's website, www.uncwil.edu/people/pricej/teaching/jobs.htm. This website includes links to job search engines, sample resumes and cover letters, a list of sociological skill sets, questions to ask on interviews, questions interviewer's may ask, job profiles, salary and compensation guides, the *Occupational Outlook Handbook* published by the *U.S. Bureau of Labor Statistics*, and information on what employers want today.

Writing Resumes

Resumes are tricky. They come in a variety styles and forms, including hard copy and **electronic resumes**. To get started with your resume, look for examples of resumes in books at your local bookstore or library and on the Internet. Also, ask your friends and mentors for copies of their resumes.

Suffice it to say that in a diverse job market, when it comes to resumes, "one size does not fit all"! You will likely want to include sections such as Professional Objectives, Educational Background, Professional Experience, Communication Skills, Interpersonal Skills, Analytical Skills, Computer Skills, Professional Activities, and Community Involvement. Under each section, list bulleted, brief descriptions of applicable information. Now what kind of information should you put under each section? Under Professional Objectives, list one or two general career goals, such as "provide assistance to children with disabilities." The "skills" sections should include specific abilities such as, respectively, writing and giving presenting reports, understanding diversity, project design and management, and word processing. We discuss skill sets with more detail in the next chapter. Professional Activities include things like conferences attended, presentations made, memberships in professional organizations, and awards received. Under Community Involvement, list any volunteer experiences you participated in, even **service-learning** activities associated with a high school or college course. If you lack professional experience, move your volunteer experiences to that category.

Speaking of professional experience, ideally, this section should be the largest section of your resume. Here, for each position, identify the job title, employer, and years of employment. Then use bulleted statements to identify the key components of each position. List phrases starting with a verb, not a noun. For example, state what you learned, developed, coordinated, applied, or accomplished at each job. Draw on job descriptions from job advertisements and informational interviews to identify the correct terminology to use. Ask yourself, if you never worked at each organization, what would be missing from it now? Remember, it takes several drafts to create a polished resume.

What sections you use and what you include in them depends on the job for which you apply. In other words, you need to tailor your resume to each individual job or position for which you apply. To facilitate this, modularize your resume. Put each section in a separate computer file, and include all of the possible content in them. Add to these files as you gain more skills and experiences. Then when you apply for a job or position, choose the sections and content from the modularized files that most appropriately reflect the skills necessary for that job. Then save this unique resume as a separate file. Again, use job descriptions that you've collected to determine what you should include in any one particular resume. Before you send it out, ask your friends, mentors, and professors to review it, and revise as needed. You may also want to create a **master resume** and ask your friends, mentors, and professors to review it too.

Writing Cover Letters

A **cover letter** represents your opportunity to guide the reader through your resume. Ideally, it should supplement or complement your resume in places that you wish to provide more detail than a resume allows. A cover letter substitutes for your resume among employers and interviewers who will not take the time to read a resume. Ultimately, a cover letter should tell the reader why he or she should consider you for a position.

Compose a cover letter from a minimum of three paragraphs. The first paragraph should identify you, your credentials and qualifications, the position for which you wish to apply, and your interest in the position. The second (and subsequent paragraphs if needed) should identify what you bring to the organization, including a description of your relevant experience and accomplishments. To identify relevant job skills and roles, turn to the job descriptions that you collected when following the job finding strategies. The last paragraph of a cover letter serves as a summary of the preceding paragraphs. It should briefly repeat your interest in the position and what makes you an excellent and, if possible, a unique candidate for the position. Finally, the last paragraph should also identify how to reach you.

Remember, the person or group (many organizations appointment a group of people to review and select candidates for positions) reviewing your application will likely evaluate a large number of other cover letters and resumes for the position. Hence, your cover letter needs to read easily and clearly, but stand out with pertinent and persuasive information. This doesn't mean you need to write a long resume. It means you need to write a well-constructed resume. Find out, from the job finding strategies we discussed above, what distinguishes competence from excellence in this job and field. Write a cover letter and resume that allows your excellence to shine.

Job Interviewing

A **job interview** is a special case of a human interaction. An interview is "inter," between two persons; hence, you have something to say about how it will evolve. What impression do you want to leave with your interviewers? You will be on stage during the entire interview process, including before the interview in your resume and cover letter, while arranging the interview, and then after the interview during follow-up communication such as requests for more information and, possibly, a job offer.

Of course, you will also be on stage during the interview. Here you want to make the best impression possible. Some obvious suggestions include: bring extra resumes and, if applicable, professional cards, and don't smoke, smell like smoke, or lie. More difficult preparation includes learning as much about the organization, the job, and your interviewers; preparing for interviewers' questions; and preparing questions to ask interviewers. Informational interviews provide invaluable insight here, as do previous work, volunteer, and internship experiences. Below we briefly summarize questions to expect and questions to ask on job interviews. For more examples, we recommend that you see Mobley et al.'s (1998) *Getting a Head Start on Your Career as an Applied Sociologist*, and the American Sociological Association's (1998) *Embarking Upon a Career with an Undergraduate Degree in Sociology.*

Questions to Expect Interviewer's to Ask	Questions to ask Interviewers
What are your short and long term career goals?	What kinds of projects is the person in the position (or the department) working on now?
Why do you want to leave your job?	What responsibilities does the position hold?
Why did you leave your last job?	Who will the person in this position work with and where?
Please provide an example of how you handled a difficult situation at work.	Why is the job available?
Please provide an example of your teamwork in one of your previous positions.	How will the new hire be introduced to the job?
Please provide an example of your individual work in one of your previous positions.	How is the person in the position evaluated? Who does the evaluation?
What are your professional strengths?	What would be success for the person in this position in one year and five years?
What are your professional weaknesses?	What support and computer resources are available to this position?
Why do you think you are a good candidate for this position?	What professional development opportunities does the organization provide?

An interview is also your time to gather information about the job, coworkers, and the organization to determine whether you would want the position, should it be offered. Try to talk to some of the employees. Do they like it here? Do they like their jobs? How long have they worked here? Try to determine as much as you can about the corporate culture. What are the collective beliefs, values, and norms in the culture? Similarly, who are the cultural heroes and what did they do? What is the norm for the use of time and space in this setting? For example, what kind of offices do people work in and how do they decorate them? How do people dress? How much interaction between employees do you observe? How do people in this setting or organization deal with deference and demeanor? The answers to these questions, and others like them, will better enable you to make a sound decision, should the organization extend you a job offer.

Negotiating Job Offers

Job offers a situation in which you may find yourself in the present and the future. In addition to answers to the questions above, you will want to ask more questions about compensation such as, "What is the starting salary?" What benefit package does the employer offer? How are raises determined? How often? Are they based on cost of living and/or merit?

About asking for a higher salary: Before you do so, consider what kind of organization extended the job offer. Government organizations, and often those affiliated with the government, follow strict policies specifying the pay range for a particular job title and the educational attainment and previous experience of the job candidate. So there will likely be little room for them to increase the salary offer. If you can find out the pay range, you may be able to negotiate up to, but usually not over, the mid-point of the range.

Similarly, non-profit organizations have little room to negotiate job salary, but for different reasons than government organizations. Non-profit organizations usually operate with tight, "shoestring" budgets, with few real opportunities to increase their financial resources. So, though they may want to offer you a higher salary, they may not be able to. However, all is not lost. In exchange for lower salaries, government and non-profit jobs usually offer higher job security, more flexible work hours, and attractive benefit packages including sick and personal days, health care plan choices, pension and other employer paid retirement plans, 401 and other pre-tax retirement savings, and child care services. Most business organizations offer benefit packages too, just maybe not as full or as flexible.

TASKS

1. Obtain a filing folder or cabinet. Create a separate file for -- Job Opportunities, Networking, Letters of Reference, Cover Letters, Resumes, Informational Interview Questions, Interviewer Questions, Interviewee Questions, and Job Offers. Now begin building the appropriate folders by filing in any information or documents that you currently have on each topic. Start adding additional information and documents to each folder as they develop.

2. Identify your career goals. What information would you like to know about this career? What are three specific ways you could find more information about this career? What are three specific ways that you could find a job in this field?

QUICK START

In this chapter, you will learn,

- The value of meaningful work.
- How to identify applicable job titles.
- How to identify your skill sets.
- How to get involved in your community.

TERMS

Analytical Skills	A person's abilities to design, collect, evaluate, and interpret data.
Communication Skills	A person's abilities to deliver and distribute information.
Computer Skills	A person's abilities to utilize computer software and hardware.
Human Capital	Skills, training, and resources that accrue within humans that may be used and/or invested in interaction with others.
Interpersonal Skills	A person's abilities in interacting with others.
Meaningful Work	Used by Marx, this term refers to the idea that work that should reward financially, personally, and sociologically.
Organizational Skills	A person's abilities to efficiently and effectively set and attain goals.
Personal Skills Inventory	A list of a person's skill sets, often included in a resume.
Political Skills Inventory	A list of a person's opinions on key social issues and the social change organizations they subsequently support.
Skill Sets	A person's abilities arranged into substantive categories, such as interpersonal skills, organizational skills, analytical skills, communication skills, and computer skills.

TOPICS

We've made the case that the nature of work is changing drastically. One thing that hasn't changed is that most of us need a job to provide for our families and ourselves. Hence, work becomes central in our lives, especially among Americans. In the years to come, you will spend more time at work than you will at home, so it makes sense to try to work at something you enjoy, something you find valuable. Take a minute to think about work that offers meaning and value to you and to society. The balance between financial reward and personal satisfaction is an important one, and we suggest that you consider this as you apply for a job. Some jobs reward financially, personally, and sociologically. Not everyone attains this career magnificence, though. Most jobs reward financially, or personally, or sociologically -- not all three. What kind of jobs will help you and others? How can you find them? Let's take a look.

Making a Difference

Armed with an inventory of personal and professional tools, how can you make a difference through **meaningful work** for you and society? First, let's consider what it might mean to "make a difference."

"Making a difference" refers to making a change or an improvement. When does a difference need to be made? Sociologically, there are at least three conditions. First, from a functionalist view, we need a difference when there is a gap between what a system provides and what it needs to provide. Second, from a conflict view, when people use power to exploit, oppress, or alienate others, we need to make a difference. Third and finally, from an interactionist view, we may need a difference any time people collectively believe that a change is needed. Making a difference may range from a revolution on one end to small incremental changes in our daily lives on the other end. Creating social change, even small daily life change, is tough. You will need to bring your sociological skills, and your personal passion, to the problem. Remember, a difference means change, and any sociologist will tell you that change is not always welcome.

In what venues can you make a difference? We can make a difference at a micro, meso or macro level. That is, respectively, in your personal space, in your family, work place or community, and in society. Sometimes people pay sociologists handsomely for our "difference-making" activities, and sometimes we offer our services pro bono, as a form of volunteering. Most compensation that sociologists receive for changing social behavior falls somewhere in between.

Now, let's look at some specific real-life examples of making a difference:

- The local school board says there isn't enough money for new textbooks this year. As a member of the PTA, you organize a systematic effort to understand the problem and develop an alternative way to fund the books.

- During his orientation to a new position in the human resources department, your new coworker notices that this department uses inefficient procedures in processing new employees. For example, it asks new hires to write down the same information on four different forms. As your protègè, he shares his observation with you. After bringing the problem to your boss's attention, you and the new coworker conduct a departmental analysis of the process and recommend streamlining it.

- The local youth athletics organization needs coaches. You conduct a needs assessment and program evaluation, uncovering new ways to recruit and compensate coaches.

Well, you might be saying to yourself, that all sounds good, but how do I find jobs that offer meaningful work for pay, with benefits? Here again, we need to translate our sociological skills into job titles. You may find this difficult at first. Most beginning job searchers do. Use the job titles list below, taken from real job advertisements, to help you get started.

109

Job Titles for Job Seekers with Sociological Skills	
Areas of Meaningful Work	*Possible Job Titles*
Program Development, Coordination, and Evaluation	Program/Project Coordinator, Program Director, Program Officer, Program Manager, Evaluation, Assistant Director, Program Assistant, Development Officer, Outreach Coordinator, Executive Director, Evaluation Officer, Regional Evaluation Specialist, Program Evaluation Specialist, Program Evaluator
Public, Social and Health Services	Community Specialist, Adoption Specialist, Habilitation Specialist, Educational Development Technician, Family Services Specialist, Assessment Specialist, Community Support Specialist, Human Services Habilitation Coordinator, Substance Abuse Counselor, Residential Counselor, Counselor, Developmental Disabilities Professional, Group Home Director, Foster Care Coordinator, Administrative Community Development Planner, Activity Director, Prevention Intervention Specialist, Quality Assurance Manager, Quality Management Coordinator, Volunteer Coordinator, Volunteer Program Manager, Grant Outreach Coordinator, Service Coordinator, Crisis Intervention Coordinator
Educational Organizations	Quality Enhancement Consultant, Developmental Disabilities Professional, Rehabilitation Instructor, Development Assistant, Head Start Director, Family Services Coordinator, Education Coordinator, Admissions Counselor, Registrar, Assistant Registrar, Assistant Director of Leadership and Volunteer Services
Recreation and Events Coordination	Athletic Coordinator, Recreation Coordinator, Recreation Assistant, Activity Director, Recreation Coordinator, Events Coordinator, Recreational Therapist, Special Events Coordinator
Civic and Non-Profit Fundraising	Planned Giving Director, Fundraising Director, Director of Planning and Development, Assistant Planning Director, Director of Public Relations/Marketing, Fundraising Consultants, Planned Giving Officer, Manager of Major Donor Programs, Manager Annual Giving Programs
Communications/Media	Internet Coordinator, Communication - Technology Director, Web Developer, Creative Writer, Producer
Criminal Justice and Adolescence	Crime Prevention Specialist, Teen Court Coordinator, High Risk Intervention Specialist, Youth at Risk Coordinator, Private Investigator, Children's Counselor, Juvenile Counselor, Probation Counselor, Community Services Juvenile Restitution Program Coordinator, Trial Court Administrator

Activism/Politics/Public Policy	Political Organizer, Fellow, Fiscal and Social Policy Analyst, Policy Analyst
Religious Organizations	Music Director, Ministry Education Coordinator, Family Life Director
Research/Consulting	Field Interviewer, Research and Information Analyst, Data Analyst, Research Analyst, Social Researcher, Health Analyst, Criminal Justice Analyst, Research Assistant, Research Associate, Data Processing Supervisor, Researcher, Community Monitoring Research Associate, Social Research Associate I, Social Research Associate II, Social Research Assistant I, Social Research Assistant II, Executive Office Research-Assessor, Research Manager, Data Quality Coordinator
Business Organizations (Note, on the surface, it often appears that few opportunities exist in business to make a difference. However, sometimes, as part of a traditional business position, you can create opportunities to improve social relations within these organizations, and improve the organization's community outreach programs. Plus, they often pay well!)	Consumer Lender, Client Services Representative, Public Relations Specialist, Public Relations Representative, Services Representative, Sales Representative, Account Executive, Compliance Coordinator, Human Resources Specialist, Recruiter, Human Resource Representative, Human Resource Assistant, Staffing Specialist, Human Resource Manager, Personnel Specialist, Relocation Coordinator, Personnel Management Specialist, Human Resource Program Facilitator, Benefits Specialist, Resource Development, Planning and Development Officer, Assistant Planning Director, Public Relations Representative, Marketing Specialist, Capacity Building Coordinator, Operations Assistant, Development Associate, Business Management: Administrator, Director of Internet Sales, Resident Property Adjuster, Contracts Specialist, Town Administrator, Property Manager, District Director, Town Manager, Management Trainee, Assistant County Manager, Assistant Association Manager, Beach Management Coordinator

Keep in mind when you are looking at job advertisements that sometimes the job title will not convey the job description. You need to read the job description to determine if you are interested in and qualified for the position. The job titles above are examples. Look for ones like them, but also keep your eyes open to new ones. Job titles for similar job descriptions vary dramatically by occupation and organization.

In addition to a paid job, consider volunteering and becoming active in your community. You can find volunteering opportunities in your local newspaper and by listening to the local radio stations. Again, don't look for a "Wanted: Volunteer with Sociological Skills" sign. Instead, look and listen for activities and events that sound meaningful to you. Then call and ask how you can get involved. Send the advertiser a letter, outlining your skills and then follow up. Often times representatives of non-profit and civic organizations are overworked and lack appropriate support resources such as clerical and administrative assistance. Don't be offended if you do not hear back from them. Instead, call them. Be persistent. This is a great way to meet new people and make a difference in your community.

Finally, you can also engage in meaningful work by participating in your local government and being an active citizen. Make it your job to understand prominent new stories of local, regional and national interest. For example, evaluate the arguments before and against education vouchers. Your sociological skills provide a unique understanding of these dynamics. Share your perspective with your friends, coworkers, neighbors, and family members. This not only further develops your sociological imagination, but also enables other people to better understand the world around them.

TOOLS

As you take more courses and collect professional experience through jobs, internships, volunteering, and community involvement, you will develop and accumulate **human capital**. At some point, hopefully sooner rather than later, you may want to take stock in what human capital you have to offer. Human capital resources fall into at least five **skill sets** that you should include on your resumes and cover letters: **interpersonal skills**, **organizational skills, analytical skills, communication skills**, and **computer skills**. We'll help you identify those skills that you gain in sociology courses, but we encourage you to think across all disciplines. Let's start.

What interpersonal skills do you possess? There are many ways to identify your personal skills. Relatively easy methods include: (1) reviewing materials from your coursework such as syllabi and tables of contents from textbooks, (2) examining job descriptions from job advertisements, (3) conducting informational interviews, (4) assessing other students' resumes, and (5) reading job profiles. Using this methodology, we classify the skill sets a sociology major develops in the table below. Some of these skills may work for your resume. If so, modify the names of the skills, as necessary, to fit your resume and cover letter. Add to the table below as you develop or identify new skills.

Interpersonal Skills	Organizational Skills	Analytical Skills	Communication Skills	Computer Skills
Team building	Group cooperation	Problem conceptualization	Writing reports	Word processing
Respect of diversity	Organizational development	Research design	Presenting reports	Spreadsheets (Excel)
Social influences on behavior	Problem solving	Surveying	Grant writing	Data analysis (SPSS)
Facilitating social change	Project design	Personal interviews	email	PowerPoint
Conflict management	Project development	Program evaluation	Web page development	Web navigation
Conflict resolution	Project coordination	Needs assessments	Searching scientific literature	Database searching
	Project management	Focus groups	Synthesizing scientific literature	Database design
	Fund raising	Data collection, analysis, and interpretation		

In reviewing your responses to the table above, you may see some skills that you both prefer and excel at applying. Reword these skills for inclusion on your resume. You may also see some skills that you prefer to use, but apply poorly. These are skills you need to improve. Finally, you may see skills that you excel at applying, but prefer not to use. These too are skills you need to improve.

TASKS

1. Create a Professional Experience section for your resume. Identify all of your professional experiences including jobs, internships, volunteering, service learning, community involvement, and so on. Don't leave out anything, whether paid or unpaid. (Babysitting, yard work, and newspaper routes all count!) With each job, identify:

   - What skills you learned or developed in doing the job.
   - What you accomplished, or what products you created, in doing the job.
   - What you left behind: If you had never done that job, what would that organization or employer not have now?

2. Start conducting a **personal skills inventory**.

- Identify all of your professional experiences including jobs, internships, volunteering, service learning, community involvement, and so on. Don't leave out anything, paid or unpaid. (Babysitting, yard work, and newspaper routes all count!) Identify for each job:
  - What interpersonal skills you learned or developed.
  - What organizational skills you learned or developed.
  - What analytical skills you learned or developed.
  - What communication skills you learned or developed.
  - What computer skills you learned or developed.
  - Add the skills identified above to your resume.

- Identify all the college courses you have taken. With each course, identify:
  - What interpersonal skills you learned or developed.
  - What organizational skills you learned or developed.
  - What analytical skills you learned or developed.
  - What communication skills you learned or developed.
  - What computer skills you learned or developed.
  - Add the skills identified above to your resume.

- Now identify the jobs for which you qualify:
  - Transform the inventories above into work roles.
  - What jobs match these work roles?
  - What job titles match these jobs?
  - How would you find these jobs?
  - Use your answers to the above questions to write professional objective statements for each copy of your resume.

- Start conducting a **political skills inventory**.
  - What local or national organizations match your social and political causes or interests?
  - How could you get involved with these organizations?
  - Write a letter of interest to send to the director of each organization.
  - Call each organization and inquire about their activities and volunteer needs.
  - Get involved!